U. R. Uzakova
S. S. Rakhimkhodjaev

Development of technology for production of new structuresjacquard tapes

U. R. Uzakova
S. S. Rakhimkhodjaev

Development of technology for production of new structuresjacquard tapes

ScienciaScripts

Cover image: www.ingimage.com

This book is a translation from the original published under ISBN 978-620-7-48433-1.

Publisher:
Sciencia Scripts
is a trademark of
Dodo Books Indian Ocean Ltd. and OmniScriptum S.R.L publishing group

120 High Road, East Finchley, London, N2 9ED, United Kingdom
Str. Armeneasca 28/1, office 1, Chisinau MD-2012, Republic of Moldova, Europe
Managing Directors: Ieva Konstantinova, Victoria Ursu
info@omniscriptum.com

Printed at: see last page
ISBN: 978-620-8-36986-6

Annotation

Contents

The work is devoted to the development of technology for the production of new structures of jacquard ribbons. Computer design of the machine causes mainly the expansion of the range of products produced and more convenient storage of information and weave pattern, and the artist-designer can on the monitor without working out samples to apply a multivariation approach to the production of this pattern, apply different combinations of tones and weaves and transfer the programme into the computer disk controlling the electronic jacquard machine, which would require a lot of time and large material costs in the development of the pattern in the classical way. Jacquard ribbons for epaulettes have been developed, taking into account new ornamental compositions and colour solutions in accordance with symbolic meanings between them. New systems of jacquard ribbons pattern coding are proposed, taking into account the weaves of ground and patterned yarns, the type of jacquard machine and the sequence of used weaves. The technological sequence of designing, preparation and production of special purpose jacquard ribbons is developed, the method of fixing sagging patterned weft threads on the wrong side of the ribbon and enhancing the effect of relief on the front side of the jacquard ribbon is specified. It is expedient to produce special purpose jacquard ribbons on the basis of local raw materials (natural silk), as this type of raw materials is available in the required quantity and price and has ergonomic properties. The formula of weft processing for two-weft jacquard fabrics with a large pattern rapport is obtained. The filling density on the base, width and number of ends in the filling when producing jacquard ribbons are substantiated. The technological process of jacquard ribbon production on the weaving machine by means of the mathematical method of rototable planning of the experiment of the second order is investigated. The geometrical interpretation of the mathematical model is studied by means of slices.

The optimum technological parameters of jacquard ribbon production have been determined. The level of warp thread breakage is 0.05 breaks per 1m of fabric at warp tension of 20 cN, the value of scoring-15 mm, the position of the scalp above the sternum by 25 mm. In the assortment of jacquard weave tapes the priority parameters are appearance, comfort, breathability, non-marring, breaking load and resistance to wet processing. In terms of physicomechanical and hygienic properties new samples of jacquard tapes meet the requirements of the Republican and European standards for the group of special purpose jacquard tapes.

Keywords: yarn, warp, weft, fabric, rapport, weave, parameters, jacquard, desint, workmanship, density, breathability, properties, tension, pattern.

GENERAL CHARACTERISATION OF WORK

Analysing the literary sources devoted to military attributes, the system of computer-aided design of fabrics, techniques and technology of jacquard weaving it is determined that many domestic and foreign researchers were engaged in this direction. They provide the solution of a number of problems related to the technology of design and production of jacquard weaving, application of CAD elements, programming and patterning in the production process, etc. Along with this, there were no scientifically grounded researches aimed at the development of design and technology of special purpose jacquard ribbons production, taking into account specific features and national traditions in ornamentation and symbolism. Therefore, research works in the direction of development of technology of production of new structures of jacquard ribbons of special purpose for raw structures of the Republic of Uzbekistan taking into account the use of national ornament is one of the actual tasks facing the textile industry.

Purpose of the research. The aim of the present work is to substantiate the concept on the development of military attributes of the Armed Forces of the RU and to develop the design and technology of production of special purpose jacquard ribbons.

Objectives of the study:

-Develop a design and automatic design system (CAD) for the production of paraphernalia for law enforcement agencies;

- justify technological parameters and process of jacquard tapes production for special purposes;
- to improve the theoretical calculation of relief weft weaving for two-strand jacquard fabrics;
- to determine consumer and physical and mechanical properties of special purpose jacquard tapes.

Object and subject of the research. Attributes of the Armed Forces of the RU (epaulettes), jacquard ribbons, methodology of their design and technology of their production.

Research methods. To solve the set tasks in the work we used methods of designing jacquard fabrics with the help of computers, methods of mathematical statistics and planning of experiment, methods of research of technological and consumer properties of jacquard tapes.

The main points put forward for defence:

- justification and recommendation of distinction marks of power structures of RU;

- development of military paraphernalia (epaulettes) for the Armed Forces of the RU;
- Proposal of a new coding system for jacquard fabrics and a programme for entering weaves on a jacquard complex;
- development of analytical calculation of relief weft weaving for two-stitch jacquard fabrics with large weft rapport;
- A new technology for the design and production of jacquard ribbons;
- Recommendation of optimal technological parameters for jacquard ribbon production on a ribbon weaving machine;
- determination of physical, mechanical and consumer properties of jacquard ribbons (shoulder straps).

Scientific Novelty:

- the distinction marks of the power structures of the RU were substantiated and proposed;
- military attributes of the Armed Forces of the RU were developed (patent application No. SAP 00570, No. SAP 00571, No. SAP 00591, No. SAP 00491);
- for the first time a new coding system and a programme for entering weaves on a Jacquard complex were proposed;
- The theoretical formula of relief weft processing for double-knit jacquard fabrics with large weft rapport is recommended;
- optimal technological parameters of jacquard ribbon production were determined;
- consumer and physical-mechanical properties of jacquard tapes were investigated and determined;
- The method of fixing the sagging weft pattern yarns on the wrong side of the jacquard ribbon and obtaining the relief effect on the front side of the jacquard ribbon is improved.

Practical value of the work. Samples of ribbons of VS attributes are developed and implemented. The possibility of production of jacquard ribbons with a given pattern and combination of weave with the help of computer modelling is substantiated. The use of this technology reduces the time of pattern processing, which increases labour productivity. Besides, the methodology of designing and calculation of jacquard ribbons is applied in the educational process when studying the design and structure of fabrics in the subject "TZLU" at the department "Technology of textile materials".

- **duplication of results.** On the subject of work 5 scientific papers have been published in the journals of the Republic of Uzbekistan. 4 patents for inventions were obtained (patent No. SAP 00570, No. SAP 00571, No. SAP 00591, No. SAP 00491).

Content of work

The introduction substantiates the relevance of the topic, formulates the purpose and objectives of the study, shows the scientific novelty and practical significance, research methodology.

The first chapter provides a review of literature sources devoted to the origin and essence of military paraphernalia, the issues of automatic design and structure of fabrics and the development of weaving technique and technology. Military paraphernalia is a complex of distinctive signs designed to manifest the specifics of military units and military specialities, the hierarchical structure of the army, its state affiliation, historical merits and traditions of certain formations. In addition, military paraphernalia includes a visual representation of the strategic military-political concept of the state. The development of technology and design of attributes for the Armed Forces of the Republic of Uzbekistan is a necessary aspect to be taken into account when developing the system of semiotic concepts "history and modernity". Semiotics, which is the theory of signs and sign systems, defines a sign as a material (acoustic or visual) signal that contains coded information and denotes the denotation: that is, a set of essential, distinctive features of a concept or object. Signs-symbols are artificially constructed ideographic signs carrying conceptual information. They are in associative connection with the object, based on a kind of social agreement. These signs have the ability to convey in a generalised, abstracted form a whole system of concepts, and their semantic volume often exceeds the semantic volume of the word - the basic lexical-semantic unit of the language. All this testifies to the unique potential of symbolism and indicates the need for careful and thoughtful work in its use. Under Amir Temur ornamental motifs were widely used - cartouches, various kinds of signs, i.e. "knots of happiness", we can assume that the image of Amir Temur's coat of arms was the tamga of the Hulaguid khan Argun. However, under him this tamga had a local character, while Amir Temur gave it a state significance, probably wishing to emphasise thereby the connection with the Genghisids, to which he attached special importance. It is well known that the circle was a kind of protection from evil and unclean forces, and three circles formed a more stable figure - their opposition. We should not forget that the circle embodies the reflection of the solar solar symbol. The semantics of Amir Temur's coat of arms contains multifunctionality of its content, in which, in addition to the interpretation of Clavijo, could figure purely pagan definitions. This should not be surprising, because the overwhelming majority of the population of Maveraunnahr at that time were Muslims, among whom the vestiges of pagan beliefs were largely preserved. The colour symbolism on items of military paraphernalia can be based on the traditional for Central Asia semantic and symbolic meanings: blue - the colour of Eternity; blue - a sign of transition to the path of true Faith (in Sufism); green - the colour of true full Faith; red - the colour of knowledge; black - in Turkic peoples from ancient times denoted Greatness, Power and Might. There is a deeply erroneous opinion that epaulettes as an element of the military uniform allegedly derive their origin from mythical metal shoulder pads that protected the shoulders of the warrior from sabre blows. However, this is just a beautiful legend, which does not have any serious substantiation. Pogony appeared on the Russian military clothes, only with the creation of the regular army by Tsar Peter I between 1683 and 1699 as a purely practical element of clothing. Its task was to keep the shoulder straps of the heavy grenadiers' grenadier bag from slipping off the shoulder. This explains its appearance: a cloth flap with the lower end sewn into the shoulder seam of the sleeve and having a slit for a button in its upper part. The button was sewn to the shoulder of the caftan closer to the collar. The epaulet was originally attached to the left shoulder. The merits of the epaulet were quickly appreciated, and it appears on the clothes of fusilers, musketeers, in short, all those who had to carry bags of various kinds. The colour of the epaulet was red for all. It is easy to notice from the images of that time that epaulettes are absent on the shoulders of all officers, cavalrymen, artillerymen, sappers. Later on, the epaulet, depending on the needs of the time, was moved then to the right shoulder, then to the left, or disappeared altogether. Quite quickly this very noticeable element of the uniform began to be used as a decorative element of clothing. The epaulettes were used as a means of distinguishing servicemen of one regiment from servicemen of another regiment since 1762, when each regiment was fitted with epaulettes of different weave from a garus cord, i.e. only since that time the epaulettes began to fulfil the second functional task. At the same time, an attempt was made to make the epaulettes a means of distinguishing soldiers and officers, for which purpose the epaulettes were woven differently for officers and soldiers in the same regiment. The lower end of the epaulet had ends hanging down, making it somewhat similar to an epaulet. This circumstance in a number of modern editions leads the authors to the erroneous statement that it is an epaulette. However, the design of the epaulette is quite different. Types of epaulettes weave appeared so many

(each regimental commander himself determined the type of epaulettes weave) that it was impossible to memorise the type of epaulettes by regiment and to distinguish an officer from a soldier. Emperor Paul I returns epaulettes purely practical purpose - to hold the strap of the bag on the shoulder. Again the epaulet disappears from the officers' and non-commissioned officers' uniforms. However, officers and generals have an axelbant on the right shoulder, the upper part of which very much resembles a garous epaulet. The second attempt to make the epaulettes a means of distinguishing officers from soldiers was made by the emperor
Alexander I, when in 1802, during the transition to the tailcoat uniform, cloth epaulettes of pentagonal shape were introduced. Soldiers received epaulettes on both shoulders, non-commissioned officers on the right shoulder (since 1803 on both shoulders), officers on the left shoulder (the axelbant on the right shoulder remains). The colours of the epaulettes were set according to the seniority of the regiments in the following order: red; white; yellow; light crimson; turquoise; pink; light green; grey; purple; blue. Since 1807 the colour of the epaulettes was set according to the number of the regiment in the division: 1st regiment red epaulettes, 2nd regiment white, 3rd regiment yellow, 4th regiment dark green with a red edge, 5th regiment light blue. From 1809 all Guards regiments were given scarlet epaulettes without cipher. Since 1807 on the epaulettes of army regiments yellow or red cord on the epaulettes laid out the number of the division to which the regiment belongs (cipher). Soldiers and non-commissioned officers had exactly the same epaulettes. Officers' epaulettes had the same colour as the soldiers of the given regiment, but were lined on all sides with gold galloon. However, in 1807 the epaulettes were first replaced by a single epaulet, and from 1809 officers wore epaulettes on both shoulders. Epaulettes disappeared from officers' uniforms until 1854. They remain an accessory only for soldier and non-commissioned officer uniforms.

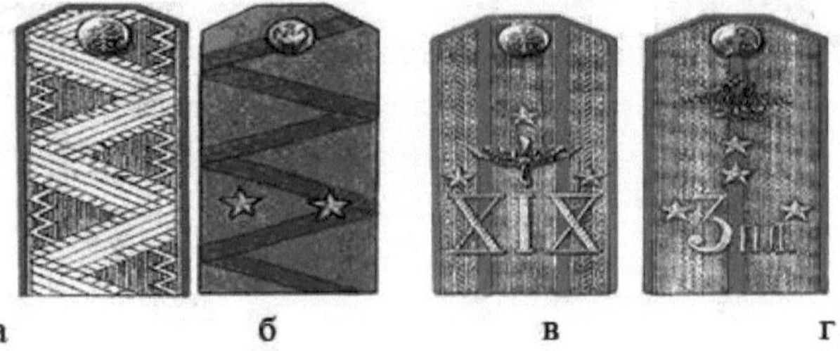

а б в г

Fig. 1. Samples of epaulettes of the Russian Tsarist Army, a- epaulettes of a lieutenant-colonel of the corps aviation unit of the 19th Army Corps, b- epaulettes of a military pilot, staff-captain of the 3rd field aviation unit, c- epaulettes of an engineer-general, d- epaulettes of a major-general of engineering troops.

Until 1843, epaulettes would carry two functional loads. Firstly, to keep the shoulder straps of the satchel on the shoulders; secondly, the epaulettes would become a determinant of a soldier's belonging to a certain division (by the number on the epaulettes) and to a certain regiment (by the colour of the epaulettes). Since 1814 all grenadier regiments in all divisions had yellow epaulettes, and the rest of the regiments of divisions: 1st regiment red epaulettes, 2nd white, 3rd light blue, 4th dark green with a red edge. Later, the colours and ciphers of the epaulettes would be changed several times (Fig.1). In 1843 epaulettes for the first time received the function of determining the ranks of non-commissioned officers. Transverse patches denoting the rank appear on them. Patches of white bason (braid) were given to infantry, jaeger and naval regiments; patches of white bason with red thread along the middle of the patch were given to grenadier and carabineer regiments. Non-commissioned officers of the nobility in all regiments had patches of gold galloon. At the same time, junkers, portupey-junkers received an epaulet covered with gold galloon. However, the same epaulet was given to subporporters and portupei-praporters. Fieldfebels had a wide gold galloon. The epaulettes, which had not been on the officers' uniforms since 1807, returned in 1854 in a new capacity. In 1854, epaulettes are first given the function of determining officer and general ranks. At this time officers and generals receive a new overcoat and galloon epaulettes on it. The epaulet was of the soldier's pattern (of the epaulet colour assigned to the regiment), on which two strips of galloon of a special pattern were sewn lengthwise for officers-in-chief so that there was a gap of 4-5 mm between the strips. On the epaulettes of staff officers one strip of wide and two strips of narrower galloon with gaps between them were sewn on. The galloon could be silver or gold (according to the colour of the instrumental metal assigned to the regiment). A strip of wide gold galloon with a zigzag pattern was sewn on the general's epaulet. The size of stars of all officers and generals was the same. The ranks of officers and generals were distinguished as follows. One lumen: ensign - 1 asterisk, lieutenant - 2 asterisks, lieutenant - 3 asterisks, staff-captain - 4 asterisks, captain - no asterisks. Two lumen: major - 2 stars, lieutenant-colonel - 3 stars, colonel - without stars. General epaulettes: major-general - 2 asterisks, lieutenant-general - 3 asterisks, general from infantry (so-called "full general") - without asterisks, general - field marshal - crossed staffs. In the period 1917 - 1943 in the Red Army epaulettes were abolished. Since 1943 epaulettes were introduced as rank insignia, which will exist until the very end of the Red (Soviet) Army in 1991 - 1993. The colours of epaulettes, their form of arrangement of stars will change, but the system of distinction of servicemen by ranks remains unchanged. In the independent young Republic of Uzbekistan, new epaulettes are introduced at the beginning of the 21st century, which have significant differences from the previous ones, taking into account national traditions in attributes. Firstly, the shape and background of epaulettes, size and shape of stars, etc. are changed. The basis of any epaulet is a woven ribbon. The ribbon can be decorated in the following ways: drawing on the ribbon by printing; drawing on the ribbon by stitching; drawing on the ribbon by weaving a group of threads (remise); drawing on the ribbon by weaving each single thread (jacquard). The variety of pattern and artistic design of epaulettes is especially revealed when using jacquard ribbons. Any weave weave is a two-dimensional set of main and weft overlaps and is depicted graphically. This way of depicting weave weaves has been around for a long time and is on the one hand sufficiently illustrative and on the other hand can be used to determine important parameters of weaving technology (e.g. the wefts in the weaving remise). However, this method is not convenient for communicating information orally, it takes a long time to depict the weave, so there is a need for an abbreviated designation of the weave. For example, the designation twill 1/3, satin 5/2, etc. shows the weave of one thread, and the way the other threads are woven is implied in the name of the weave. In our case twill means a constant shear equal to one, and satin equal to two. Another system is also used, where the amount of shift is indicated, and the weave of the first thread is hidden in indices and in the number of equations, for example: yd=x; ys= 2x; etc. Nowadays, methods of automated control of techniques and technologies and fabric design (CAD) have found wide application. At creation of CAD of fabric for automated catalogues of weave weaves, automation of images of weave weaves, necessary at drawing up of different series of documentation, in the end, input of these parameters in memory of the computer. We propose a system of designation of weaves oriented on the use of computers to represent a weave by a two-dimensional matrix, in which digit 1 denotes the main overlap, and 0 denotes the exact overlap. In such an image, the result of multiplying the punching matrix by the weave matrix represents the cardboard matrix. Direct input in the form of a matrix of ones and zeros is impractical because it requires manual preparation of the weave pattern and is time consuming. That is why nowadays various methods are used to save the input of the weave that is understandable to the dessinator and in a form convenient for practising on the

COMPUTER. With the use of a computer, a system of working out different weaves in fabrics according to a single threading pattern in the remise has been created. Fabrics of the same article but different patterns can be produced according to a single threading pattern. In this case the use of a computer allows to calculate for the production of several weaving patterns the single warp threading in the remise and thus to replace in some cases the warp threading operation by the warp tying operation. For mathematical description of the problem of calculation of single warp stitches and algorithms of its solution it is suggested to use the apparatus of the theory of relations and the listed variant of the method of branches and boundaries. The algorithms for the calculation of single warp plugs were implemented in the form of a package of applied programmes. The system "Autodessinator" was created on the basis of the computer, which allows to design crepe weave patterns for fabric production on weaving machines with eccentric and dobby shedding mechanisms. Parameters determining the fabric structure: raw material composition selected taking into account the purpose of the fabric and requirements to them; warp and weft yarn diameters and their ratios, in which an increase in the yarn diameter of a given system increases the breaking load and elongation of the fabric of this system; warp and weft densities and their ratios, in which a change in the fabric density of one yarn system causes a change in the technological parameters of fabric structure and properties; evaluation of the tension of fabric production on the loom, which is characteristic of the weaving process. The weaves with the smallest number of overlaps have a higher breaking load and yarn work in the fabric, and the weaving is accompanied by a higher tension on the loom; technological parameters, which should include warp and weft tension and their ratios, the amount of overstep, the size and location of the shed. These parameters change the arrangement of threads in the fabric, and hence the structure of the fabric, in particular the working of threads in the fabric. It should be noted that mainly the works are directed on development of CAD of fabrics of single-layer main and combined weaves of formulas and workmanship is built on the basis of geometrical model for single-layer fabrics. Jacquard fabrics are the best in terms of their artistic design, aesthetic properties and variety of assortment. However, it takes about three months of manual labour to produce a simple pattern, to patronise and make cards for a jacquard machine. Time reduction is possible in the following combination: by using CAD fabrics; by producing jacquard cards directly from patterns; by using a modern electronically controlled jacquard machine. Modern weaving machines can save energy, reduce environmental pollution, increase the degree of automation of production, operate the machine with great ease and simplicity, and use better and more modern software to produce jacquard fabrics. These include Jacquard machines of "Jakob Muller" (Switzerland), "Bonas" (Great Britain), "OMM" (Italy), "Macheba" (Germany) with electronic control and electronic pattern programming. The application of electronic control and computer-aided design in jacquard weaving is caused by mobility of assortment renewal, elimination of unproductive labour when creating a pattern and transferring it to the machine, elimination of a large volume of paper cards, reduction of raw material consumption when refilling the pattern and ensuring high quality of fabrics. In summary it should be noted: at the present stage the development of military paraphernalia items (epaulettes, chevrons, etc.) for the Armed Forces of the Republic of Uzbekistan taking into account national traditions in ornamentation and symbolism is one of the most important tasks in the political sphere of the independent Republic of Uzbekistan; the development of technology and design of paraphernalia for the Armed Forces of the Republic of Uzbekistan is a necessary aspect in which it is necessary to take into account the semantic concepts "History - modernity"; mainly the works are aimed at the development of CAD of fabrics of one and the same name.

In the second chapter design and design of attributes of power structures is given, elements of attributes of power structures are justified, modelling and design of attributes by means of computer graphics, software of system of construction of weaving weaves, computer design of attributes of power structures, new sketch of jacquard ribbons of special purpose is developed. Attributes have been accompanying the development of civilisation since ancient times. Attribute comes from the Latin word attribution and means "giving, endowing". In the fine arts, an attribute is understood as an inherent, material distinctive feature of a deity or a hero. It can be allegorical or symbolic images, examples of which can be, for example, a stick (attribute of Hercules) or scales and a blindfold (attribute of the goddess of Justice). Attributes are one of the expressions of the imperial and philosophical experience of previous generations and have retained their relevance to the present day. One of the ideological tasks facing our young independent state is the development of a new concept of attributes for the power structures of the Republic of Uzbakistan. Not all the searches undertaken in this sphere in recent years have been successful: there are examples of successful realisation of interesting ideas, but there are also many developments that testify to the authors' opportunistic considerations and their unprofessionalism. A specialist intending to work seriously in the field of creating attributes, it is necessary to know the reasons for the traditional use of the most common ornamental motifs. Semiotics, which is the theory of signs and sign systems, defines a sign as a material (acoustic or visual) signal that contains coded information and denotes the denotation: that is, a set of essential, distinctive features of a concept or object. Since ancient times, the oak tree, characterised by its strength and longevity, has been identified by all peoples with the concepts of strength, power and inflexibility. Laurel is the personification of determination and courage, thanks to which laurel wreaths were awarded to the winners who showed these qualities in various spheres of human activity: science, sports, military art, poetry, etc. Olive is a symbolic equivalent of wisdom and peacefulness, which is confirmed by the mention of this symbol in canonical religious texts, as well as in myths and legends of various peoples. Signs-symbols are artificially constructed ideographic signs carrying conceptual information. They are in associative connection with the object based on a peculiar social agreement. These signs have the ability to convey in a generalised, abstracted form a whole system of concepts and their semantic volume often exceeds the semantic volume of the word - the basic lexical-semantic unit of the language. All this testifies to the unique potential of symbolism and indicates the necessity of careful and thoughtful work in its use. The class of mythopoetic signs identical in their form to geometrical elements is of certain interest in this sense. Previously they were widely used in mythological and religious spheres, and nowadays they are used in attributes, symbolism and emblematics. They are attractive because they provide stability, accuracy and conciseness of modelling due to their relative geometric simplicity: ball, cone, pyramid, square, rhombus, combination of straight, curved and broken lines. Semantic codes created by applying these elements provide adequate symbolic representation of real objects, concepts and other latent (hidden) information. A geometrical symbol has always depicted the structure of the cosmos in its horizontal and vertical aspects (in contrast to structureless chaos, which has never been described with the help of geometrical symbols). Such symbols underpinned the organisation of ritual space and are also easily established in the forms of sacralised objects. Nowadays the most widely used geometrical lines are straight or broken line, various types of regular curves: spiral, currency correlated in religious and mythological symbolism with thunder, lightning, earth, water and the like (Fig. 2.1. a). An example of a widespread symbol of this type is the meander, originally the name of a river in Asia Minor. According to the myth, this river, characterised by its meandering and representing a continuous line broken by right angles, dried up when Phaeton's chariot approached the Earth (Fig. 2.1. b). This symbol denotes the concept of beginning and end, as well as the concept of eternity. In ancient China the meander symbolised reincarnation and thunder, in ancient Greece it symbolised the labyrinth of the legendary King Minos. In the East, the idea of the meander received its own interpretation: it was used in the ornament of the madrasah Kalan, used as a background ornament in men's clothing, was an element of decoration of different types of weapons and eventually became one of the typical forms of ornament. Of the geometric symbols, which include the circle, square, mandala, cross and swastika, the symbolic images of "regular" polygons deserve special attention: the triangle in various mythological contexts symbolises the fruit-bearing power of the earth, marriage, unity, security, as well as the conceptual triads "life - death - rebirth", "body - mind - soul". Three connected triangles are the symbol of the Absolute, the Pythagorean symbol of health, as well as the Masonic emblem (Fig.2.1c). The image of a triangle inside a square symbolises the connection of the divine and human, heavenly and earthly. Two intersecting triangles symbolise divinity, the union of fire and water, the victory of spirit over matter. The symbolical image of a square, starting from such geometrical qualities as number four, absolute equality, simplicity, order and straightness after heraldic and attributive metamorphosis begin to designate concepts of rightness, truth, justice, wisdom and honour (fig. 2.1.d). The square and the circle correlated with it can symbolise the horizontal plane of the world tree. In this case the corners and the centres of the sides denoting four or eight basic directions of an idea are given special value. In these points are located deities, trees personifying the sides of light, winds and basic, cosmic elements and elements (Fig. 2.1.e). Aztec schemes inscribed in a square, the elements of which are correlated with a special circle, serve as an example of world modelling. The square (rectangular) or octagonal scheme embodies the classification system of binary oppositions describing the world: top - bottom, right - left, good - evil, etc., as well as the idea of the basic elements of the world: fire, water, earth, air. The square structure is used not only to denote the spatial properties of the universe (sides of the world, directions, centre), but also symbolises the basic temporal coordinates: four parts of the day, four seasons, four world ages, four human ages, etc. One of the most widespread mythopoetic symbols present in modern military attributes is the square or its double sign - octagon. It is difficult to say anything unambiguously about the semantics of the octagon, as there is a considerable number of different interpretations. According to one of them, the octagon in architectural ornamentation symbolises a mandala. This is one of the main sacral symbols of Buddhist mythology, denoting a country, society, aggregate, assembly, space, orbit, ring, wheel, etc. Mandala belongs to the number of geometrical signs of complex structure. The most characteristic scheme of a mandala is an outer circle with a square inscribed in it. In this square is inscribed inner circle, the periphery of which is usually depicted as an eight-petalled lotus, or eight segments. The square is orientated on the sides of the world. In the middle of each side of the square are T-shaped gates, continuing outside the square with cruciform images. In the centre of the inner circle a sacral object of veneration - a deity, his attribute or symbol is depicted (Fig. 2.1.e). Pentagon - a regular pentagon in the form of a star symbolises eternity, perfection, the universe. In Muslims it is a combination of a star and a month; in Aztecs it is an emblem of Thoth, Quetzalcoatl, Mercury; in Christians it is a symbol of five wounds of Jesus Christ; in American Indians Pentagon is a totem sign. In the basin of the Central Asian Inter-Area the circle, rhombus, cross served as symbols or attributes of solar deities. They are often found as symbols of nature in combination with images of animals, birds, "tree of life". Sometimes the sun symbol looked like a wheel with spokes and "caps" on the rim. The likening of the sun to a wheel goes back to cosmogonic myths, which identified the circle with the solar disc. The centre of the wheel is an immeasurable point, the rim is an immeasurable circle; the inner space of the wheel contains good and evil, life and death, light and darkness, etc. (Fig. 2.1.g). In the military equipment of Amir Timur's army, the solar sign was often found. It can be found on the solar plexus^ on the armour, on headdresses. Geometric symbols often become an element of artistic form. Symbols and signs form a significant layer of mythopoetic means, the creative use of which allows to influence the psyche by modelling new situations. This property explains the attraction of symbols and signs to create emblems and paraphernalia, capable of influencing the human subconscious. Totem images also belong to the sphere of attributes. The word "totem" in the language of the Ojibwe Indians means "his clan". Totemism was widely used in the epoch of early matriarchy, was characteristic for agricultural cultures, in the epoch of patriarchy totems were used by shamans, and also were elements of the cult of ancestors. The emergence of totems is explained by the belief in the supernatural connection between humans and animals and plants. The attribution of supernatural qualities to various animals was characteristic of such religions as Zoroastrianism and Buddhism, which sought to explain in this way the manifestations of the power of nature and other factors affecting human life. In the Central Asian region, the images of bull, horse, deer, camel, ram, goat and boar, which personified the forces of nature and the cosmos, were once widely used. On the obverse, and sometimes on the reverse of silver and copper coins of Amir Timur of all denominations placed a sign of three rings, made in different compositions: with one ring at the top and two at the bottom or vice versa - with two rings at the top and one at the bottom. These rings were enclosed in a triangle, and in other cities enclosed (Yazd, Kumm) in a circle. This sign Clavijo called the coat of arms of Amir Timur. Timurbek's coat of arms is three circles - it means that he is the king of three parts of the world (Fig. 2.1 .h).

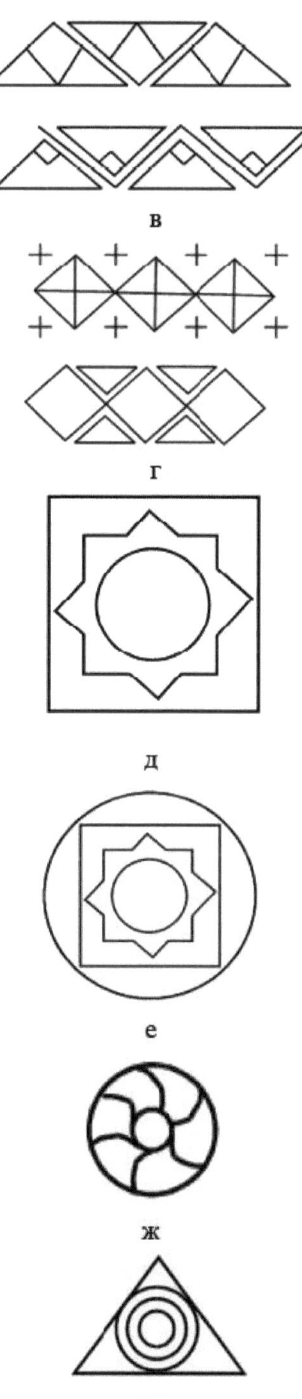
в
г
д
е
ж

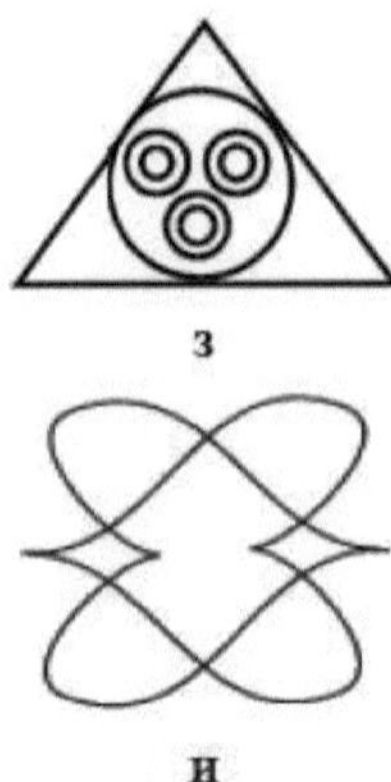

Figure 2.1. Symbolism in the attributes of the WARU.

Under Amir Timur ornamental motifs were widely used - cartouches, various signs, i.e. "knots of happiness" (Fig. 2.1.i), we can assume that the model for the coat of arms of Amir Timur was the Hulagenta tamga of Argun Khan. However, under him this tamga had a local character, while Amir Timur gave it a state significance, probably wishing to emphasise thereby the connection with the Genghisids, to which he attached special importance. It is well known that the circle was a kind of protection from evil and unclean forces, and three circles formed a more stable figure - their opposition. We should not forget that the circle embodies the reflection of the solar solar symbol. The semantics of Amir Timur's coat of arms contains multifunctionality of its content, in which, in addition to the interpretation of Clavijo, could figure purely pagan definitions. This should not be surprising, because the overwhelming majority of the population of Maveraunnahr at that time were Muslims, among whom the vestiges of pagan beliefs were largely preserved. The colour symbolism can be based on traditional for Central Asia semantic and symbolic meanings: blue - the colour of Eternity, blue - a sign of transition to the path of true Faith (in Sufism), green - the colour of true and full Faith, red - the colour of Knowledge, black - in Turkic peoples from ancient times denoted Greatness, Power and Might. On the basis of our research, taking into account national traditions in ornamentation and symbolism, using the best that was in the past era, especially in the era of Amir Temur, we propose new structures of samples - standards of attributes of the Armed Forces, the technology of their development. Competent combination of symbols of the ancient world and modern dissenatorial skill in the implementation of technology and design of attributes of power structures (epaulettes, chevrons, buttonholes, insignia) of the Armed Forces of the

Republic of Uzbekistan will highlight our arguable advantages in relation to foreign analogues. The term "design" appeared in our country recently. Before that, the design of things was called "artistic construction", and the theory of creating things "technical aesthetics". In translation from English, "design" means drawing. The word also gave rise to industrial concepts: "designer"-artist-designer, "design-form" - the external shape of the object and others. Consider design as the activity of the artist-constructor in the field of designing mass industrial products and creating on this basis subject environment. The value of each thing in two beginnings - usefulness and beauty. In every object there is a technical and aesthetic beginning, always impermanent and historically changeable. Practical usefulness of a thing does not require explanation, but it turns out that usefulness can be accompanied by some aesthetic experience. The use of computers in scientific research is a prerequisite for the study of complex systems. The traditional methodology of the relationship between theory and experiment should be supplemented by the principles of computer modelling. This new effective procedure makes it possible to study holistically the behaviour of the most complex systems, both natural and those created to test theoretical hypotheses. An effective way to overcome the difficulties of creating real samples is to build a computer model of the phenomenon under study, which is understood as a set of numerical methods for solving the basic equations, algorithms for their implementation and computer programmes. A good computer model turns a computer from an ultra-fast calculator into an intellectual tool that facilitates the discovery of new effects, phenomena and even the creation of new theories. The effectiveness of a computer model is largely determined by the quality of the software used. The main requirements for software are, of course, ease of input and correction of input data, as well as visualisation (visibility) of the results of the calculation. Today there are both powerful specialised programming systems (MAPLE, SolidWorks, AutoCAD, etc.) and special programmes in which convenient graphical user features are implemented. The use of computer models turns the computer into a universal experimental setup. The computer experiment provides full control over all parameters of the system, computer experiment is cheap and safe, with the help of computer it is possible to set "fundamentally impossible" experiments (geological processes, cosmology, ecological catastrophes, etc.). The process of modelling in AutoCAD Designer is reduced to the fact that firstly to set a typical profile on a plane, and then to give it spatial properties, having built the so-called basic form, and then to add to it new design and technological elements (standard or described by typical profiles). Creation of typical profiles of form-forming elements in AutoCAD Designer takes place in two stages (at that, the performed actions are as close as possible to the operations carried out by designers in everyday practice): first, a conceptual sketch of a profile is built on the so-called sketch plane, and then geometrical relations are imposed on its elements and parametric dimensions are entered. By default, when creating a basic form, the XY plane of the user coordinate system is used as a sketch plane, but the profiles of other design elements can be defined in planes other than the initial one. In this case you must define a new sketch plane using the AMSKPLN command (Sketch Plane option in the Parts menu, Sketch submenu or the Sketch Plane option in the Parts menu, Sketch submenu). To orient the sketch plane in space, you can use both the faces of the existing model and special non-form-forming construction elements - work planes. In addition to working planes in AutoCAD Designer for binding form-forming elements during modelling, other non-form-forming structural elements are also effective: working axis and working point. Computer-aided design allows not only to create, but also to improve a complex product, to evaluate and test it not at the real enterprise, but in the virtual reality environment. This is especially relevant for expensive, complex, unique technological and technical complexes. Also important is the appearance of products, its forms, characteristics - design. Design is a new field of application of computer graphics in industry. Usually the purpose of design development of a new product is to choose the most successful concept of the external appearance of products from a variety of options and detailed visual analysis of the selected concept. If the design of a product is carried out with the help of a computer, it allows to reduce several times the time both for design development and for the general development cycle (for example, the release on the market of such a complex product as a car can occur one to two years earlier). There are also significant cost savings, as all aspects of appearance are evaluated on computerised rather than full-scale models. The design part of the overall production cycle includes: conceptual modelling, i.e. preliminary development of several product variants, resulting in "three-dimensional sketches"; creation of computer "drawings", which are orthogonal projections of the future product (in traditional design, such drawings could be the final result of the work); modelling itself: tracing of drawings, i.e. creation of three-dimensional objects with their help, and then - construction of surfaces on these objects; evaluation of the design of the product; creation of computer "drawings", which are orthogonal projections of the future product (in traditional design, such drawings could be the final result of the work). Three-dimensional modelling is an area of functional intersection between a design system and CAD, but the purpose of modelling in these systems is different. For a designer, a three-dimensional model is just a preliminary construction, on the basis of which photo images are obtained. At the same time, it should be noted that the actual process of development of a new product occurs in the mode of close cooperation between designers and technologists and contains feedback, which allows at the stage of design development (and not with the finished product) to bring the model "up to speed", and this makes the use of computer technology vital for the future product. Thus, from the very beginning, the forms of the future object are harmonised with the requirements of designers and technologists. The object created with the help of modelling systems can be placed in different environments, simulate and trace not only its movements in the virtual space created for it, but also demonstrate its functioning. It is expedient to translate code descriptions into a weave pattern with the help of the Dessinator Interpreter (Dessint), which is a package of application programmes implementing the considered algorithms of weave construction. The classification of weave construction methods (Fig. 2.2) defines the structure and the principle of operation of the Dessint system. The construction methods in the Desint system are divided into three groups. The first group includes construction methods from initial elements such as a single thread or a group of threads. The second group includes methods of construction from initial elements of parts of pattern patterns. The third group includes special methods of construction of rapports with the arrangement of the pattern. The desired group is determined by the type of the initial element necessary for the construction of the final, desired weave. The Desint system is based on the following principles: in the process of constructing a rapport of one weave in one group, the transition to another group is excluded; in the process of constructing a rapport in a group, two rapports are involved, where at least one rapport is constructed from the initial elements; in the process of constructing a rapport in a group, corrections of completed rapports are allowed; the system is implemented as an interpreter of code descriptions. Input data files are usually punched onto magnetic tape or perforated tape and written to disc memory by a computer operator. Input files of code descriptions of small volume can be created without any difficulties by the technologist-dessinator himself from the keyboard of the display. For this purpose, firstly, it is necessary to study the rules of combining several code descriptions into a sequential file according to the operator's instruction, and secondly, it is necessary to know how to create and edit texts from the computer display. An arbitrary name is assigned to the created text file of code descriptions and it is written to the disc memory. In the system of computer-aided design (CAD) of single-layer fabrics the subsystem of weave generation should cover all known single-layer weaves and the possibility of designing new weaves by new methods of their composition on the basis of presentation of information to the computer in a procedural way is provided.

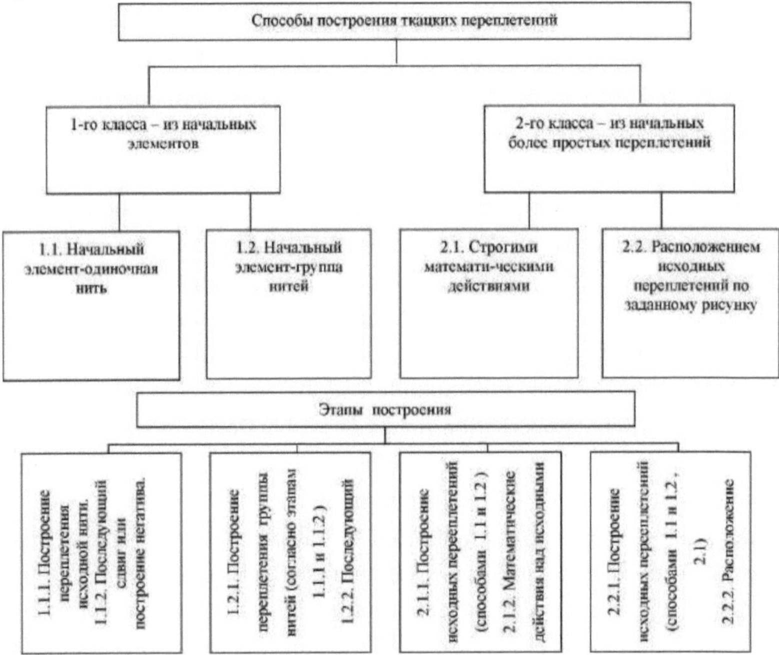

Fig. 2.2 Classification of single-layer weave construction methods.

The procedures of weaving weave composition are related to the technological peculiarities of weaving and new methods of weave composition are designed not by creating new procedures, but by changing their interrelation. CAD of fabrics opens wide opportunities for dessinators - it reduces the time for search, evaluation and selection of optimal variants of fabric structure with execution of all necessary documentation, increases the level of manufacturability of these solutions, accelerates updating of fabric assortments.

Designing with the help of CAD should be carried out in the shortest possible time with minimum expenditure of raw materials and labour, and the resulting product should be fashionable, beautiful, high quality, have the lowest possible cost and meet all the initial operational requirements. At automation of fabrics designing the experience of the dissenator to fabrics designing is preserved, but at the same time there is an opportunity by modern means of computer technology to create a "tool" of designing, which will allow dissenators and technologists to view and evaluate the results of different design solutions. Designing is provided by man-machine procedures, in the course of which labour-intensive constructions and calculations of variants are assigned to the computer, and the technologist is left to evaluate the variants and make optimal design decisions. Therefore, the purpose of fabric design is to determine the main parameters of its structure, the combination of which in the fabric at minimum consumption of selected raw materials gives the best satisfaction of all the properties arising from the purpose of the fabric. The basis of the methodology of determination of the structure parameters is the knowledge of dependences of the fabric operational properties on the structure parameters. At the solution of these tasks the information bank is formed, which is the core of information support of the automated design system. The proposed CAD of fabrics requires from the researcher to carry out purposeful scientific experiments on studying dependences of separate technological and operational properties on fabric structure parameters. The dessinator or weaving technologist should prepare a design task, which contains three components of initial data: type of raw material, required properties, weave. The dessinator has to choose: from a variety of raw materials a given type of raw materials by properties, by linear densities of warp and weft yarns and their combinations, which determine the texture and appearance of the fabric; requirements to the properties of the designed fabric, which must have a certain set of operational properties and is set by the dessinator based on the importance of this or that property for the designed fabric; weave of the fabric, which determines the manufacturability, operational properties and appearance of the fabric.

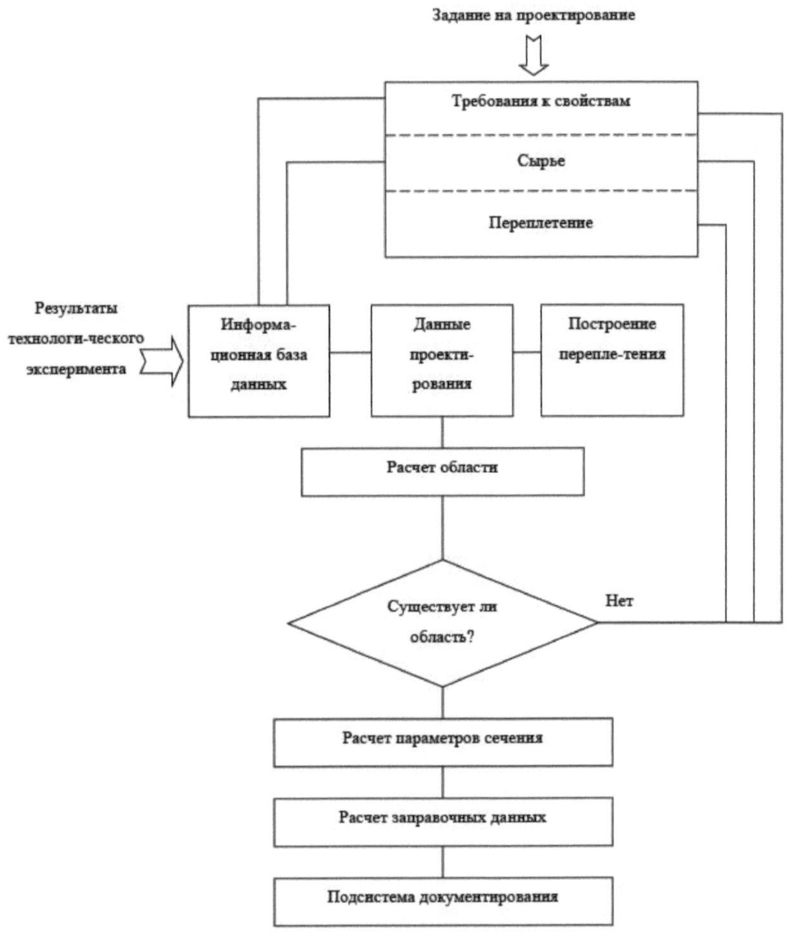

Fig 2.3. Structural and functional scheme of CAD of fabrics.

The methodology of computer-aided design implies drawing up a code description of the weave, on the basis of which the Desint subsystem builds the rapport, calculates the digital characteristics of the weave and generates the filling documentation. Fig. 2.3 shows the structural-functional scheme of the fabric CAD system. The system consists of the weave subsystem, information and calculation subsystems, which are oriented to the dialogue design mode. Data transfer between subsystems is carried out through the common design data area. The calculation subsystem is intended for calculation of the permissible design area in the coordinates specified by the dessinator, according to the parameters of the fabric structure. The design region is constructed according to the set of properties specified by the dessinator in the design task, where the dessinator selects a certain point within the allowable region, which gives the optimal values of the fabric structure parameters. The calculation of the dressing and fabric properties, calculation of the fabric cross-section parameters and printing of the design results are then carried out. Consequently, the dessinator does the following: selects the type of raw material, linear densities of warp and weft yarns; designs the weave, i.e. a code description is prepared, according to which the weave construction subsystem builds the weave; calculates the numerical characteristics (average overlap of warp and weft yarns, the coefficient of connectivity or tension of fabric production); builds a new weave in case of unsatisfactory parameters of the considered weave; makes a transition to the design subsystem with obtaining the satisfactory parameters of the weave In the absence of the acceptable area corrects the requirements to the properties or carries out additional technological research aimed at improving the fabric quality and obtaining new functional dependences of the properties on the parameters of the fabric structure. Pattern modelling is carried out by means of incomplete or oriented method of automatic fabric designing. Under the incomplete designing is understood the autonomous use of separate CAD subsystems, where the values of the coefficient of connectivity or tension of fabric production on the weaving machine are revealed, according to which the warp and weft yarn densities necessary for this weave and the accepted type of raw materials are determined. Oriented design involves the development of a fabric prototype. Construction of the design area in the coordinates weft density - fabric production tension allows to receive new fabric samples without weaving machine threading by selection of weave and weft density. Fig. 2.4 shows the algorithm of weave construction from simpler weaves.

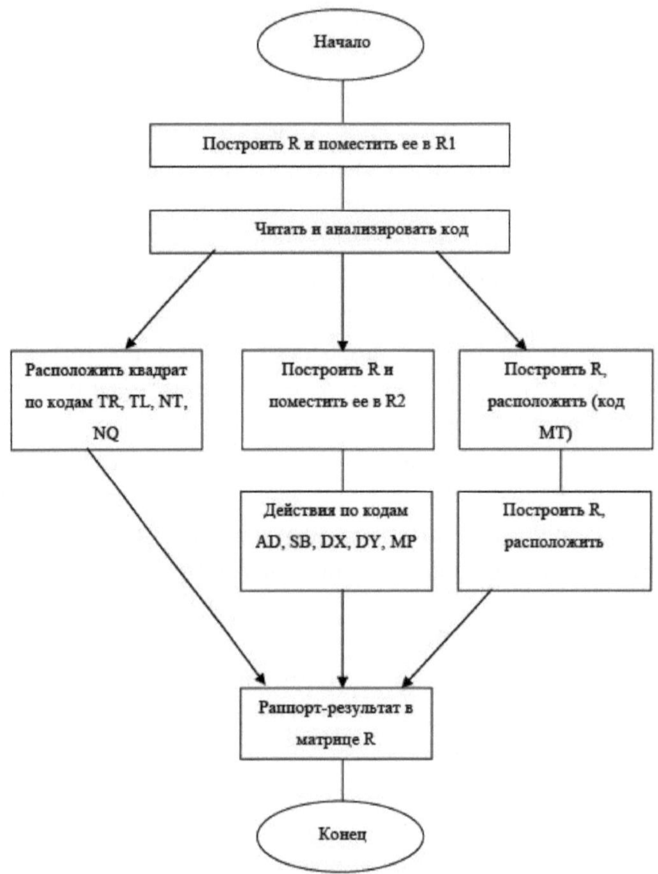

Fig.2.4. Scheme of weave construction from simpler weaves.

Making weaves according to a motif requires the compilation of a motif matrix and two initial weaves. Unlike a motif, a pattern is represented by a special matrix, which is encoded by the dessinator. The elements of this matrix can be any symbols familiar to the user. All arrangements with transformations are characterised by the fact that they require only one initial weave, which is a square. Two complex weave construction methods should be mentioned separately: the pattern arrangement and the twill arrangement. The coding of a pattern arrangement is characterised by a pattern matrix, with two, three or more initial matrices required. To perform the arrangement by pattern, it is necessary to provide the input of the pattern, determine the required number of initial matrices to fill the pattern, and then perform the arrangement of each matrix in the same way as in the case of the arrangement by motif. The structure of the code description of woven twills differs considerably from all the others. The elements of this matrix can be any symbols familiar to the user. All arrangements with transformations are characterised by the fact that they require only one initial weave, representing a square. Two complex weave construction methods should be mentioned separately: the pattern arrangement and the twill arrangement. Pattern arrangement coding is characterised by a pattern matrix, with two, three or more initial matrices required. In order to perform the arrangement by pattern, it is necessary to provide for the input of the pattern, to determine from it the required number of initial matrices to fill the pattern, and then to perform the arrangement of each matrix in the same way as in the case of the arrangement by motif. The code description of woven twills differs considerably from all other twills in its structure. The creation of the required assortment is possible on the basis of multivariant and variation solutions for the design of fabrics and products. Computer-aided design allows not only to create, but also to improve a complex product, to evaluate and test it not at a real enterprise, but in virtual reality. This is especially relevant for expensive, complex, unique technological complexes. The method of computer-aided design, compilation of code descriptions of weaves on a computer-based dissenatorial interpretation is proposed. The coding system - a model of the process of composing weaves - is an integral factor in designing fabric design. Considering the disadvantages of existing coding systems - the main disadvantage of which is unacceptability to jacquard weaving, a new coding system is proposed, taking into account the weaves of ground warp and coloured weft yarns. In the proposed new coding system, a precise argumentation is given for the type of Jacquard machine, for the sequence of weaves used. The new coding system "Palette of weaving weaves" is introduced as a subprogramme in the existing computer programme "MüCAD". The existing weaves are grouped together (based on the main weave group) and labelled in the following ways: plain weaves with the letter T (taffeta); satin, satin weaves with the letter S; twill weaves with the letter D. For example: decoding of the code system NT 624 T 1 and NT 624 S8 1 will be NT - longitudinally cut label tapes, shuttleless rapier weaving machine; 624 - total number of hooks; T - ground - taffeta; S8 - ground - 8-thread satin; 1 - number of labels in the weave rapport. In order to create a pattern, it is necessary to have a so-called pattern header: work hook plan; pattern name; weave palette; weave weave and colour preset (BV). The BV plan is provided for up to 12 weft yarn colours, including ground weft. The input menu provides an overview of all basic colours (P), each with a range of up to seven composite (shade) colours and weave weaves for pattern, ground and fray ties, as well as floating lengths and shifted weave points. The weave weaves for top, warp and ground can be changed individually for each main colour and its corresponding composite colours. It is possible to shift the individual weave points on the upper and lower border of the pattern part (top weave) to the left or right side. Table 2.1 shows the pre-sampling for the ground weave.

Table 2.1.

Preliminary sampling of soil entanglements

№	Designation of the pre-sample	Name of ground weave
1	BVT	Taffeta
2	BVS5	5-nit atlas (BVS5RE-5-TH--nit atlas with opposite ground S5RE-carnH)
3	BVS8	8-nit atlas (BVS5RE-8-MH--NIT atlas with opposite ground S8RE-carnH)
4	BVS53	8-thread atlas with extra dots
5	BVPA	2-thread main rep
6	BVK3	Twill 2/1
7	BVK3N	Twill

Weave weaves in weave preselection plans must not be mixed with another weave. Example: Weaving weave BVS5 does not mix (fit) with BVS8 or BVT, etc. A double-hued jacquard ribbon with one warp colour and two weft colours (table 2.2.). After entering the weave data, a coloured pattern will appear on the monitor. This "F-drawing" will be saved and, if necessary, can be used for corrections, amendments. The drawing to be transferred to the loom, the so-called P-drawing, will be stored separately with the initial command "A-5" by clicking the frame [* Save "P-drawing" ready for transfer to loom] *) and calling the name of the drawing to be recorded.

Table 2.2.

Two-hole jacquard tape

Plan of the working hooks	KLG:352
Picture name	KLGTEST
Weave palette	KLG
Pre-sampling of weave weaves	BVLG (EL41,EL42,TL43)

Number of primary colours	1
Reading in the direction of the base	(the number of base colours across the width is equal to the expansion ratio)
Reading in weft direction	2 (number of weft colours equals the expansion factor)

Only this "P" pattern can be transferred to the weaving machine via MULOAD. Order of operation. Change the data in the following sequence, completing each change with [INTER]: a) Name of the work hook plan, pattern class; b) Name of the pattern head: CAMHEAD (do not use other names for this purpose); c) Name of the pattern head: CAMHEAD (do not use other names for this purpose); d) Name of the work hook plan: CAMHEAD (do not use other names for this purpose).

Weave palette; d) Pre-set weave selection; e) Number of main colours (without priming); f) Number of main threads per cm (for taffeta - 57); g) Number of weft threads per cm (for taffeta - 25); h) Width and height of the sketch in mm. The position of the beginning of the pattern to be scanned, measured from the starting line j) Click [Save] to design, confirm this new "Camhead" pattern. Otherwise, the borrowed "Camhead" or drawing will be overwritten. The width [step h)] is the actual width of the sketch in mm excluding the edges and the retaining main threads (tack threads) on the left and right > Work Hook Plan. The MUCOMP 3 is designed for fast, highly accurate and extremely reliable scanning of colour photographs, colour drawings or sketches. This device allows you to write drawing programmes on the MUCOMP 3 system with a high degree of automation and with maximum productivity. After scanning, all the colours of the standard BV plan are shown. It is possible to add tinted, mixed and double colours by selecting from a section. The MUCOMP 3 T system is used to quickly and easily create pattern programmes for electronically controlled jacquard machines using sketches and outlines of the pattern project. Features: creation of multicoloured jacquard patterns on warp and weft up to 2688 hooks and 12 weft colours; 16 Megabyte memory expandable up to 40 Megabytes; graphic controller for 256 colour shades; high expansion capacity of the CCD camera (up to 600 diopters) for reading colour patterns up to 290 x 420 mm. Storage devices and memory capacity: flexible discs (floppy disks) 3 Щ 1440 Kilobytes; hard disc 3 Щ approximately 210 Megabytes; optical discs; 2x300 Megabytes; connection plugs for external (peripheral) equipment: keyboard; 20 video monitor with RGB - signal (signal of the main colours of the image red-green-blue - red-green-blue - red-green-blue) (60 images per second); graphic tablet 290 x 420 mm with 5 - button "mouse" ; inkjet colour printing device, 300 diopters. Options: computer network operation with up to 8 parallel stations; VUDESIGNER module with Corel Draw; ON LINE connection (system mode) with a separate ARTWORK station (photo-original); equipment for programming of foreign Jacquard machines and cardboard-section machines. Each document can be visualised, and changed on the screen by one or more defined dialogues. Textile defined documents are used to convert a design into a textile product ("Pdesign") and finally into a Jacquard file that can be sent to the weaving machine. Each document can either be a "root"-document (at the top of the hierarchy) or a "sub-document" and thus part of the "main" document. Typically, all the textile information needed to convert a design into a Jacquard file is contained in a pattern plan, which is a sub-document relative to the design. One of the main conceptual differences between MüCAD and its predecessors, into which the MüCAD Müller system is copied, is the Mücomp ZT, which weaves in MüCAD weave plans and palettes as copies of the original pattern files. The textile translation of the project is made on the basis of local copies, (sub-documents), which makes it easy to transfer the complete project with all its textile information from one computer and user to another by only transferring the project document file. All necessary information for textile translation is contained as sub-documents in the project document (pattern plan, drawing plan, weave plan). A new sketch for special purpose ribbons has been developed. The special purpose jacquard ribbons, worked out on the Jakob Muller machine, are designed for epaulettes of senior and junior officers and have the following semantic code meanings of these patterns. Figure 2.5 shows samples of epaulettes of senior officers. The designs are presented in three colour variations: ceremonial - golden background; shirt - white background; everyday - mugwort. The senior officer staff, is distinguished from the junior officer staff by two longitudinal distinctive stripes, which, in accordance with the branches of service, have their own special colour variations. In the proposed new design programme of special purpose ribbons we proceeded from purely - specific semantic interpretations inherent in national and historical expressions, taking as a basis the era of Amir Timur. The composition of the design of senior officers has the following semantic meaning: straight lines are interpreted as correctness and clarity of the chosen path; inscribed squares as honesty, intelligence and order of thought. The square belongs to the solar signs, which symbolises order, honesty, justice, wisdom and honour. The square structure is used not only to signify the spatial properties of the universe, but also symbolises the basic time coordinates: the four parts of the day; the four seasons; the four world ages; and the four human ages. Figure 2.6 shows samples of epaulettes for junior officers. The composition presented for junior officers has the following semantic meaning: the circle represents society, goodness, and the country as a whole. In the Timurid army, the circle was a protection from evil and unclean spirits, and three circles personified the opposition. On the coat of arms of Amir Timur there was a motif depicted in the form of three rings.

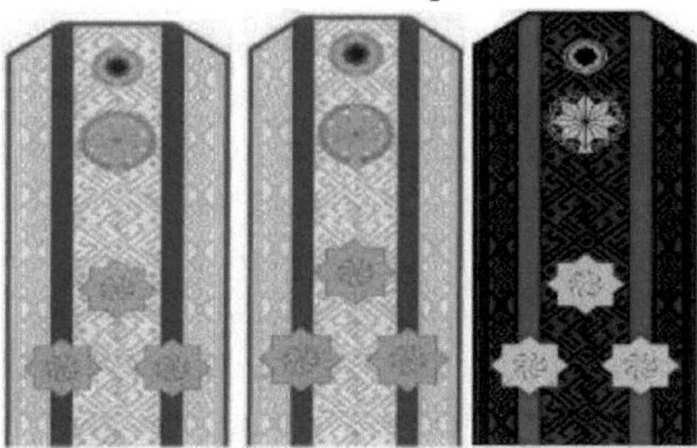

Fig.2.5 Samples of shoulder straps of senior officers

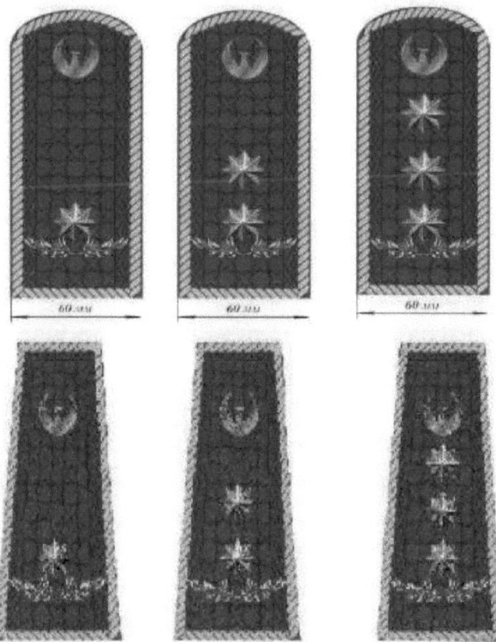

Fig. 2.6 Patterns of epaulettes for junior officers

Rhombus as well as square is a solar sign and carries the following semantic code: equality of all; purity and order of thought; correct structure of society and correctness of ideas. In combination with a circle rhombus can be interpreted as a mandala. Mandala is a sacred sign of society. Mandala belongs to the number of geometrical signs of complex structure. The most characteristic scheme of mandala is an outer circle with a square inscribed in it. On the edges of the composition of epaulettes for junior officers there are cartouches. In the Timurid army they were interpreted as knots of happiness. The paper presents variants of samples of epaulettes for junior officers. The technological chain of production of polyester jacquard special-purpose ribbons is developed for designing special-purpose ribbons. The choice of raw materials is carried out taking into account the possibilities of jacquard equipment. In the summary the elements of modern attributes of power structures of the Republic of Uzbekistan, taking into account national traditions in ornamentation and symbolism of Amir Timur's epoch, are justified. Computer design of the machine causes mainly the expansion of the range of products produced and more convenient storage of information and weave pattern. An artist-dissenter can on the monitor without producing samples apply a multivariation approach to the production of a given pattern, apply different combinations of tones and weaves and transfer the programme to the computer disc controlling the electronic jacquard machine. This would have been time-consuming and costly if the pattern had been developed in the classical way. Jacquard ribbons for epaulettes have been developed, taking into account new ornamental compositions and colour solutions in accordance with symbolic meanings between them. New systems of jacquard ribbons pattern coding are proposed, taking into account the weaves of ground and patterned yarns, the type of jacquard machine and the sequence of used weaves. Jacquard ribbons are designed with the new MuCad software, which has unlimited fabric rapport possibilities and high speed of fabric design.

The third chapter contains technological research of Jacquard ribbons, technology of Jacquard ribbons production, selection and research of Jacquard ribbon weaves, optimisation of Jacquard ribbons production. Ribbon formation is the process of interweaving of two systems of yarns at joint action of weaving machine mechanisms performing technological operations: warp tension and optusk, shedding, weft laying in the shed, weft yarn surfing to the down and winding of the fabric. In this way, the fabric formation process takes place all the way from the warp to the market roll with the mandatory participation of all the mechanisms of the weaving machine. On a Swiss Jakob Muller sliver weaving machine with a Jacquard head, the sliver formation process takes place as follows (Fig. 3.1).

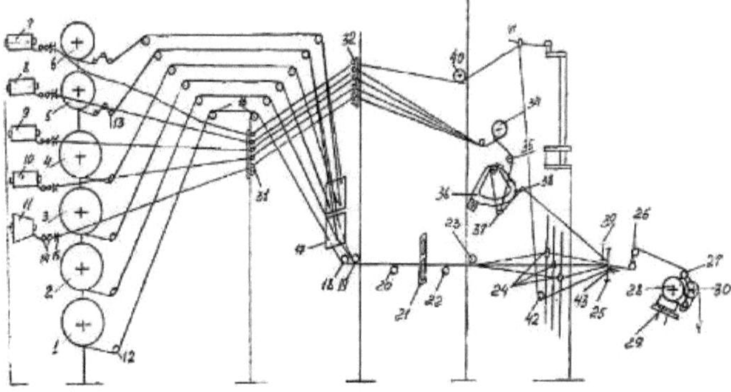

Fig. 3.1. Technological scheme of the Jacob Mullar weaving machine.

The warp yarns are unwound from the weaving heads 1-6, which are mounted on the fixed frame in two vertical rows, under a certain filling tension F. Of the twelve weaving pillars mounted on the frame, four pillars 5-6 have edge yarns and eight pillars 1-4 have background main yarns. Two navoys with background warp yarns and one edge navoy are involved in the formation of one sliver. The warp yarns envelope a system of guide rollers 12,13,16. The yarns from the two background yarns and one edge yarn are fed into the rear reed 17. Then the threads go round the guides 18,19,20, enter the lamellar warp monitor 21, into the eyes of the gale 24 of the Jacquard machine, forming a shed into which the weft threads 8, 9, 10, 11 are laid with the help of a needle, then it is nailed to the tape edge by the reed 25. The weft yarns (7 - edge dressing yarn, 8 - ground weft yarn, 9, 10, 11 - patterned weft yarns) are wound from the packs 7, 8, 9, 10, 11 located on the movable creel. They pass through tensioning devices 14, thread guides 15, 31, 32, 33. Further, the edge dressing thread 7 passes through the control device 41, guides 42 and is fed into the thread feeder of the reed needle 43. The reed needle is fixed on a rod which moves reciprocatingly from the cam. The ground and pattern threads go round the thread guides 33,34,35 and pass through the control device for monitoring the weft threads 36, through the eyes of the Jacquard machine, the separating rows 39 with the help of the needle are laid in the shed at the ribbon edge. The goods regulator withdraws the worked up sliver from the working area of the machine, and the warp is fed to form a new sliver element. The cycle is then repeated and the next fabric elements are formed. The worked sliver wraps around the guide rolls 26, 27, the roll 28, the thermo regulator 29, the guide rolls 30, 31 and is wound on the goods roller 32. When the patterned weft is laid, no sliver withdrawal takes place because the roll does not receive movement, according to the programme of the Jacquard machine. Regardless of which warp is installed on the weaving machine, the most important condition for a correct weaving process is to create the necessary filling tension of the warp and to ensure that the warp is tempered according to its consumption during the formation of each fabric element. Consequently, as the fabric is produced and taken off the loom, the corresponding warp length must be supplied, which is made up of the length of the produced fabric and the warp allowance. When weaving machines are equipped with main brakes, the warp is turned and hence the warp is released through an elastic filling system - fabric and warp tension. As the fabric is worked up, the fabric and warp tension increases and as soon as the moment of inequality between the warp tension and the braking force of the warp is reached, the warp starts to turn and the warp is released into the working area of the machine. On a Swiss-made Jakob Muller Jacquard Jacquard head sliver weaving machine the main friction brakes are installed. As the yields of the background and selvedge yarns are different, so are the designs of the main brakes. On the background main brake, the warp yarns 1, having a tension F, are wrapped around the movable shaft 4, the pre-tension of the brake belt 5 wrapped around the warp pulley is T (Fig. 3.2). The desired warp tension is created by loading or unloading the spring 2. The spring acts on the movable arm 6, on which the guide shaft 4 is embedded. One end of the brake belt 5 is fixed to the movable arm 6, and the other end of the belt is fixed to the stationary arm. As soon as the warp winding system is activated, the warp thread tension 1 causes the guide shaft 4 to be lifted, as a result of which the warp is released from the brake belt 5 and the warp is winded. As soon as the warp winding system stops, the warp guide 4 is released downwards by the tension of the spring 2. The brake belt 5 is in contact with the surface of the warp 3, hence the warp winding stops. By changing the tension of spring 2, the brake belts 5 are pressed with greater or lesser force against the pulley of the warp 3. The required warp release resistance is created between the belt and the warp, and therefore the required filling tension of the warp. Maximum braking torque $M_{Max} = TR_t (1- e^{-/ a})$, where *a is the* angle of the brake belt around the braking pulley of the navoi; R_t *is the* radius of the braking flange of the navoi; */ is the* friction coefficient; *e is the* base of the natural logarithm.

Equation of equilibrium of the lever 6 in the break-free mode *Faisin β+Taisin* β_1 - P/a_2 , where *ai and* a_2 *are the* design parameters of the brake; *β and* β_1 *are the* angles of inclination of the warp coming off the warp; *P is the* force of the spring 2.

A value of T < 0 indicates that the brake is not operating without interruption.

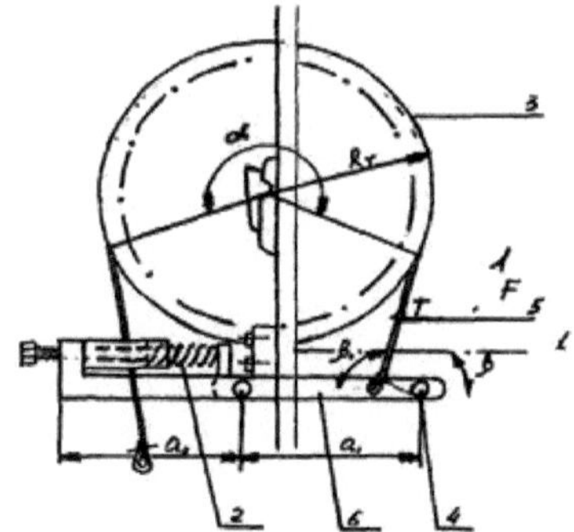

Figure 3.2. Main brake of background filaments

On the edge main brake, the thread tension is set by tightening or releasing bolt 2 (Fig. 3.3). When the thread winding mechanism is activated, the selvedge threads 1 circle the shaft 4 downwards. As a result, the small flanges of the weaving warp 6 are released from the friction force of the support 5 and the thread is twisted. As soon as the thread winding mechanism stops, the tension of the yarns 1 on the shaft 4 is no longer applied and the shaft 4 returns to its original position.

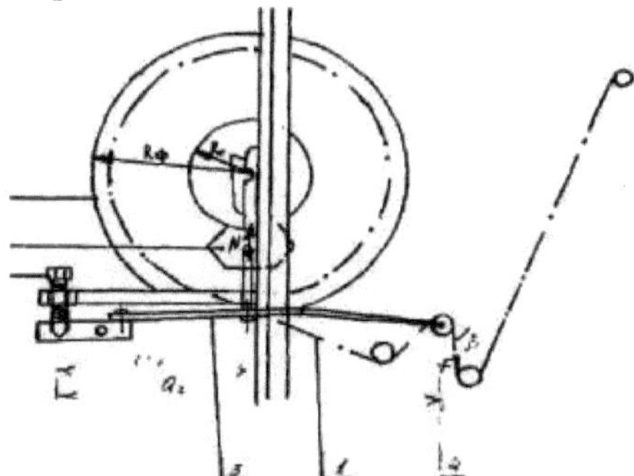

Figure 3.3. Basic edge thread brake

Consequently, the metal plate 3 is lifted upwards, the necessary resistance to *the* release of the filling is created between the support 5 *and the* flange of the filling, and consequently, the filament is stopped. A contract between the support 5 *and the* braking surface of the warp flange is a prerequisite for the operation of the edge yarn brake. *The* maximum braking torque *Mmax=fNRt,* where *f-coefficient of* friction; *N-pressure* of the lever on the warp; *Rt* - radius of the braking flange of the warp. Equation of equilibrium of the brake lever 3 in the breakage-free mode *Nai- Fa2 sinβ - Pa2* , where , *α2* - design parameters of the brakes; *β* - angle of inclination of the warp coming off the warp; *P* - displacement of the bolt. The breakage-free mode is broken if $N< 0$. In the mechanism of weft laying and selvedge formation (Fig.3.4), the weft is laid in the shed with the help of needle 1, which performs progressive-return movement. In the zone of hook 2 of needle 1 gets thread 4, lifted by galleys 7 of the Jacquard machine and is laid in the shed, and thread 3 is not fed into the zone of hook 2 and therefore does not participate in the formation of the weave pattern. On the side opposite to the needle movement, the selvedge is worked with the help of tongue needles 5, where the loop is knitted from additional (backstitch) thread 6, wound from a special bobbin. The Swiss-made "Jakob Muller" Jacquard-head tape-weaving machine, which is one of the most advanced technologies of the world textile machinery, has a computer "Mucomp 3 Junior", equipped with an optical-ellectronic scanning device. This frees the artist-dessinator from the necessity of making cartridges and speeds up the process of pattern processing. The use of a computer allows us to do without making samples, as on the monitor we can see the drawing in the weave with the specified colour shades. Fig. 3.5 shows the technological chain of designing special purpose jacquard ribbons. The design stage from pattern development to transfer of the weave pattern programme was discussed in the previous chapter.

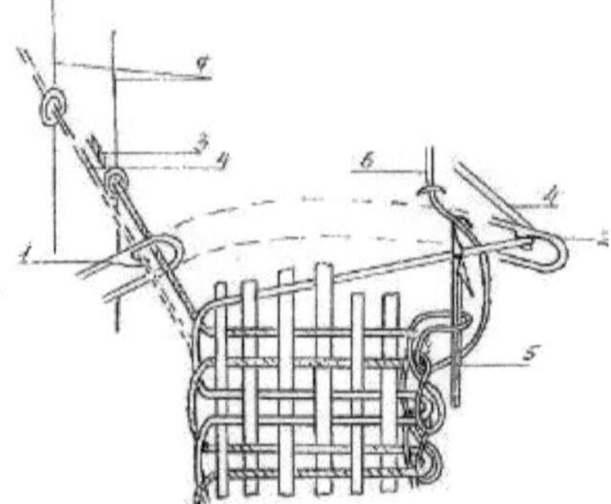

Figure 3.4. Weft insertion and selvedge formation

In the technology stage we prepare warp and weft for weaving, we weave the jacquard tape on the weaving machine, develop and finish the semi-finished product to the finished product (epaulet). Table 3.1 and 3.2 summarise the characteristics and parameters for the weaving and weaving equipment. And table 3.3 shows the Jacquard tape filling chart for warp and weft.

Here is the technical calculation of the jacquard ribbon.

1. Background. 1 background - polyester thread 12,6 tex, 600kr./m count 176 threads. 2 background - polyester thread 12,6 tex, 600kr./m count 176 threads. 3 edge - polyester thread 12,6 tex, 600cr./m count 48 threads, Total 400 threads.
2. Weft. 1 background - polyester thread 8,7 tex, 120kr./m., 2 pattern - viscose 16,6 tex, 120kr./m. 3 edge - polyester thread 5,7 tex.
3. Belt width $B_л$ =69 mm.
4. Base yarn density P_o =60 n/cm.
5. Weft density P_y =50 n/cm.
6. Rear reed felling, a) background 1 and background 2 - two threads per reed tooth, b) edge - two threads per reed tooth.
7. Picking in the front reed *6n + bnh Zraz + 6n + 4 x88raz + 6n.*
8. The reed number is 100 teeth/dm.
9. The picking in the arcata is row by row.
10. Fill width across the reed. In_3 *=71,2 mm.*
11. Coefficient of relatedness

$C = RoRu * T_{ep} / (1000 * F) =60 *25x2 * 12.6/(1000 * 10.08) = 3.6.$

$T_{ep} = 12.6+8.7+16.6/3=12.6.$

$A=2\ K_Щ K_H\ .\ (e + e_{щн}) =2 * 28 *18.\ (28+72) = 106086$ $K_щ$ *and* $K_н$ taken from Fig. Zu16u

12. Surface and linear density of the tape,

a) Weight of warp and weft yarns per 100 m of sliver, g,

$m\ o = no *To * Kyp /10 = 400 * 12.6 * 1.05 /10= 529$ *g,*

$m\ y = Ty *B *P_{ЛУ} *K_{УР} /10 = 8.7* 6.9 *25x2 *1.08 /10 = 324$ *g,*

$m_{узор} = T*K_{УР} /10 = 16.6 * 323.3 /10 = 536$ *g ,*

б) Linear density of tape, g/m

$P_c = m_o /100 + (m_y + m_{узор}) /100,$

$P_c = 529/100 + (324 + 536)/100 = 13.89$ *g/m.*

13. Tape surface density , g/m ,^{2}

$P = P_c /Vl = 13.89 / 0.069 = 201.3$ *g/m .*^{2}

14. Ribbon Fill Factor,

a) Linear filling of the tape on the base

$Z_o = do *Po = 0.139 * 600 = 83.4$ *%.*

*Do = 0.0316 C Jr = 0.0316 * l.24jdJß = 0.139mm.*

б) Linear filling of the sliver in the weft

*Pe= de *Pe = 0/138 * 50 * 10 = 69 %*

*de = 0/0316 * 1/24 ftp =0/116*

*de = 0/0316 * 1/24 JJß = 0/160*

de = 0/116 + 0/160 \ 2 = 0/138 w/

в) Surface filling of the tape

$Зп = З\ о + Зу - З_о * Зу/100,$

$Зп = 83{,}4 + 69 - 83{,}4 * 69/100 = 94{,}8\ \%.$

Ribbon Filling,

$H_{cp} = H\ T_{oo} + HyTy/(T_o + Tu),$

$N_{cp} = 0.95 * 12.6 + 0.72 * 12.65/(12.6 + 12.65) = 0.83\ ,$

$H_o = P_O / Pomax = 60/63 = 0.95,$

$Nu = Ru/Ru\ max = 50 / 68{,}6 = 0{,}72.$

$P_o = 100 * Ro / (doRo + dyKy) = 100 * 28/(28 * 0.139 + 0.138 * 4) = 63$

$Ky = n_y / Ry = 72/18 = 4.$

$Py\ max = 100 * Ry / dyRy\ doKo) = 100 * 18/(0.138 * 18 + 0.139 * 2) = 68.6.$

$Ko = no / Ro = 56/28 = 2$

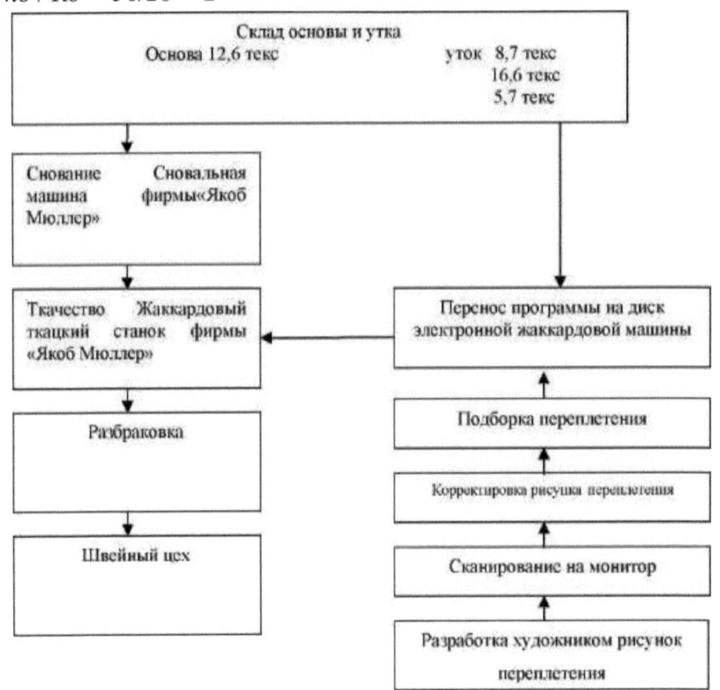

Fig.3.5: Process chain and design of jacquard ribbons.

Table 3.1.

Parameters of the Jakob Muller Jacquard weaving machine for the production of special-purpose Jacquard belts

Elements of characterisation	Indicators
- Main shaft speed, min'[1]	390
- Warp tension, cN.	20
- Weft yarn tension, cN.	15
- Number of produced tapes, pcs.	4
- Height of the shed along the reed, mm.	30
- The size of the shoulder, mm.	15
- Width of produced belts, mm.	60
- Number (colour) of weft in one belt, pcs.	5
- Yawing mechanism Weft density, yarn/cm.	Jacquard machine
- Method of edging on the side opposite to the needle	50
starting point	knitting with extra yarn 12
- Number of navoi, pcs.	

Table 3.2.

Characteristics and parameters of the Jakob Muller rewinding machine.

Elements of characterisation	Indicators
- Linear weaving speed, m/min.	50-500
- Knitting method	parallel winding 0.7-0.8 10
- Winding density on the filling, g/cm^3	224,5
- Weight of washers, gr.	102
- Dimensions of navoi, mm.	255
Small navy:	304
flange diameter	102
barrel diameter	275
flange spacing	400 washer
Big navy:	1225
flange diameter	1170 -
barrel diameter	2400
flange spacing	6000 -
- Creel:	3000
number of working bundles	10170
tension type	
- Overall dimensions of the machine, mm.	
width	
depth	
height	
- Overall dimensions of the bobbin, mm.	
width	
length	
height	
- Distance between machine and creel, mm.	

-	Total installation length, mm.	

Table 3.3.

Jacquard ribbon threading card

Basis

№	Name	Raw materials used	Linear yarn density, tex	Number of threads	Work rate	Quantity of raw materials for production of 100 m of sliver, gr.
1	Von	Polyester yarns	12,6	176	1,05	232,848
2	Von	Polyester yarns	12,6	176	1,05	232,848
3	Edging	Polyester yarns	12,6	48	1,04	62,899
4	Total			400		528,595

Duck

№	Name	Raw materials used	Linear yarn density, tex	Ribbon density, threads per 1 cm.	Coefficient of labour efficiency	Quantity of raw materials for production of 100 m of sliver, gr.
1	Von	Polyester yarns	8,7	50	1,08	324,162
2	Pattern	Viscose threads	16,6		323,3	536,678
3	Edging	Polyester yarns	5,7		12,8	7,296
	Total					1396,731

Modern Jacquard weaving machines are standardised, i.e. with several filling densities, it is possible to diversify the range of Jacquard fabrics. A small difference in filling density in the warp is compensated for by the weft density. On the same type of machine with the same jacquard machines, the same number of working hooks is selected and the same filling width is adopted regardless of the filling density. Jakob Muller sliver weaving machines have a standard filling and a warp density of 60 threads per 1 cm. Consequently, standardised threading systems must take the following nuances into account. The choice of fabric filling width depends on the filling width of the machine. The number of hooks working affects the size of the purl, the size of the pattern on the warp and the choice of weave for the background. The number of working hooks divided by the filling width must

be a whole number, and this whole number determines the number of rapports or parts in the filling and the filling density shift of the fabric on the warp. If an odd number is obtained, the number of working hooks or the filling width is changed. In all calculations, the number of hooks and the filling width of the background are taken into account without taking the selvedges into account. Jacquard fabrics can be simple (single layer), where no more than two thread systems are used, and complex (multilayer), where at least three thread systems or more are used. For epaulettes we use two and three weft jacquard fabrics, in which one warp and two or three weft thread systems are involved in the formation of the ribbon, and the weft threads are arranged in two or three layers in the ribbon. These fabrics are characterised by the fact that the weft threads necessary for the pattern are allocated on the front side of the ribbon, and in the threads where they should not be allocated, these threads are removed to the underside of the fabric. The peculiarity of the production of these tapes is that the goods regulator diverts the fabric when laying the background (ground) weft and does not divert the fabric from the forming zone when laying relief (patterned) wefts. This causes the embossed wefts to be packed on top of the background wefts in layers rather than side by side. Consequently, when determining the fabric (tape) density it is necessary to have: the number of systems of main and weft yarns; the density of yarns in each system; the alternation of yarns of each system forming one ground. Table 3.4 shows the results of calculations on selection of warp density (P_o), width of one filling part (*4)* and number of filling parts *(N)* for filling density (standard) on warp equal to 60 threads/cm, number of working hooks 352 and total width of sliver filling *Sh=25.2 cm.* Data definition and calculations were carried out according to the formulas. The number of parts per threading *(N)* and the shift *(C) of* the density on the warp. *N = C = K / W=352/25.2=14,* therefore the number of parts *N=14* pieces and density shift on the basis *C=14* threads. The width of one piece in the filling (W*) W = K / W = 352/196=1.8 cm.* Maximum filling density of warp yarns *(P).* P_o *= K/4 = 352/1.8=196 threads/cm.* To determine the subsequent filling densities, subtract the density shift (*C)* from the previous density.
Po2 = Pol - C; Po3 = Po2 - C; $P_{o4} = P_o h - C$; etc. Determination of each variant width of one part of filling in variants is necessary to divide the number of working hooks (*K)* by the filling density of variants, i.e. $41 = K/P_{ol}$; $Ch_2 = K/P_o h$; $Ch_3 = K/P_{03}$; etc.
To determine in variants the number of filling parts (*N)* it is necessary to divide the total filling width (*W)* by variants of width of one filling part, i.e. *N*

$= W/Ch, N = W/Ch_2 ; N = W/Ch_2$; *etc.*
For the machine "Jakob Muller" from all variants (Table 3.4) the most acceptable threading and working out of the tape of the 11th variant, as the machine is set to work out four ends of jacquard ribbons 11th variant: $Po = 56$ threads / cm; $K = 352$ hooks ; width of one part of the threading $4 = 6.3$ *cm -, the* number of parts in the threading $N = 4$, *the* total width of threading (without edges) $W = 25.2$ cm. The total number of background and selvedge hooks is 400 hooks plus hooks for changing the weft colour and disconnecting the goods regulator when laying the patterned weft, the number of hooks will be about 416 hooks in total. Hence, if the Jacquard machine has 640 hooks, there will be only 416 working hooks (Fig.3.6). In terms of composition, patterns for jacquard ribbon designs can be different - geometric, symmetrical, striped, plaid, with a border repeating a certain rhythm of details, or arbitrary arrangement, free composition.

Table 3.4.

Results of calculation of basis density, width and number of tapes in a filling

№	Number of working hooks, *K*, pcs	Filling density of fabric on warp, *P*, yarn/cm.	Width of one part of the filling, *Ch*, cm,	Number of refuelling parts, *N*	Total filling width, *W*, (without edges) cm.
1	352	196	1,8	14	25,2
2	352	182	1,9	13	25,2
3	352	168	2,1	12	25,2
4	352	154	2,3	11	25,2
5	352	140	2,5	10	25,2
6	352	126	2,8	9	25,2
7	352	112	3,1	8	25,2
8	352	98	3,6	7	25,2
9	352	84	4,2	6	25,2
10	352	70	5,0	5	25,2
11	352	56	6,3	4	25,2
12	352	42	8,4	3	25,2
13	352	28	12,6	2	25,2

WORK-HOOK PLAN:				H416S8
(WEFT DESIGN)				
MACHINE				640
No. OF HOOKS TOTAL				416
No. OF	L:	8	R:	8

HOOKS SELFEDGE				
No. OF HOOKS HOLD-ENDS	L:	4	R:	4
No. OF HOOKS DESING				392
FIRST HOOK DESINGN SELFEDGE				1
WEAVE, SELF-EDGE:				HHK1.3
HOLD ENDS FIGURE:				R2.2
SPECIAL: HLD. ENDS TOGR..:				HF8
SPECIAL: HOLD ENDS FIG. :				HF4N
WEAVE, HOLD ENDS GROUND:				G8
WEAVE, GROUND:				S8N
SAVEPRINT				OUT

Fig.3.6. Parameters of the designed special-purpose belt.

Fig.3.7 shows the palette of weaves entered into the computer for designing patterns on jacquard ribbon for epaulettes. As can be seen, various combinations of weaves and their individual use allows you to significantly diversify the pattern (pattern) on the front side of the jacquard ribbon (epaulettes). Due to the fact that polyester threads are imported from abroad (in particular, from France) and the cost of this raw material is very high, taking into account transport costs, costs for shade colours, etc. we propose to replace this raw material (polyester) with threads of local raw materials, i.e. threads of natural silk. In this case we have developed and produced a pilot batch of a new assortment of jacquard ribbons for epaulettes with natural silk threads of 15.7 tex and 13.75 tex.

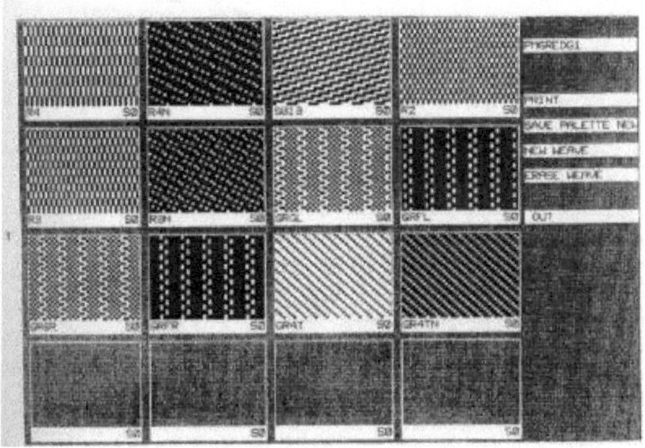

Figure 3.7. Weave palette for jacquard ribbon

Since silk yarns have mercerisation and bulkiness inherent to it, the appearance of ribbons turned out to be much better in terms of aesthetics than epaulettes made of polyester yarns. According to the technological indicators, jacquard ribbons made of natural silk have lower values of the relief weft processing coefficient, which allows to reduce the consumption of raw materials in the production of jacquard ribbons for epaulettes. Besides, from the sanitary and hygienic point of view, installation of ribbon edge melters on some types of machines leads to emission of gases and odours, which cause tearing and suffocation of workers in the process of melting polyester and viscose weft yarns. Therefore, it is advisable to replace chemical fibres in the weft with natural fibres (natural silk). Table 3.5 shows the main technological parameters of jacquard ribbons. Here it is necessary to remember that preparation and production of jacquard ribbon for natural silk and polyester yarns is the same, i.e. remains unchanged according to Fig. 3.5.

Table3.5.

Jacquard ribbon production rates

Indicator	Raw materials	
	PEF 8.7 tex x 2	Natural silk 13.75 tex
Belt width, mm	68 + 2	68 + 2
Ribbon weft density, yarn/cm.	50	50
Linear density of one linear metre of tape, g/m.	18 + 0,9	17,9 + 0,9
Intertwining	Jacquard	
Number of threads in the warp	400	400

Coefficient of workmanship ground weft 1 background 2 background relief weft 1 pattern 2 pattern	1,08 1,07 367,57 367,57	1,09 1,09 347,45 342,32
Yarn consumption per 100 m, gr.	1843	1800

Cross-sections of jacquard ribbons (shoulder straps) made of viscose ribbon patterned weft yarns and natural silk weft yarns were also made. The analysis of the slices shows that patterned weft yarns of natural silk have more volume than patterned weft yarns of viscose fibre. Consequently, the appearance of the ribbon has a more pronounced pattern due to the relief (convexity) of the threads in the weave of the fabric. The arrangement of patterned threads on the underside of the ribbon without fixing leads to the formation of long planking, which later must be cut off (converted into ugars) or transferred to the epaulet product, which complicates the technology of epaulet manufacturing. In any case, the epaulettes are not form-stable, due to the thickness of the tape. In order to obtain form-resistant epaulettes and to reduce the technology of ribbon processing, we propose fixing sagging pattern threads on the wrong side of the fabric. In this case, in order to prevent the visibility of this weft on the front surface of the fabric, it is placed under the overlaps of the background weft or under the overlaps of the long deck of another patterned weft. A palette for fixing weft yarns on the wrong side of the tape (Fig. 3.15) was developed and entered into the computer. Therefore, having in the computer a palette of weaves on the front side of the tape (Fig. 3.7) and a palette of fixing of patterned weft threads on the wrong side of the tape (Fig. 3.15), the computer will calculate and set the necessary fixing of the patterned weft without breaking the pattern (colour) on the face of the fabric.

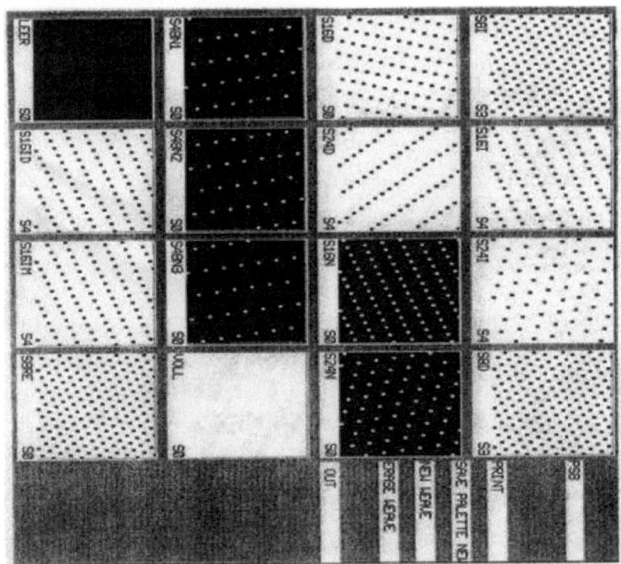

Fig. 3.15. Palette for fastening patterned weft yarns on the wrong side of the sliver

The warp and weft yields are one of the main factors that determine the structure and properties of jacquard ribbons as well as the raw material consumption. Weft yield is determined by cutting a standard (accepted) of the jacquard ribbon sample.

Workmanship on the basis of

$$a_o = \frac{l_o - l_T}{l_T} * 100\%$$

Finishing in weft

$$a_y = \frac{l_y - l_T}{l_T} * 100\%$$

where l_o, l_y - length of the straightened yarn of warp and weft respectively; *1t* - length and width of the standard sample of jacquard tape.

Table 3.6 and Table 3.7 show the experimental values of warp and weft yields for a fragment of the jacquard ribbon weave pattern (Fig. 3.16).

Table 3.6

Numerical characteristics of the yarn count in the sliver

№	Name	Indicators					
		Repetitions, *m*					Average value of *U*
		Y1	*Y2*	*Uz*	y_4	y_5	
1	1 base background	1,05	1,06	1,04	1,07	1,03	1,05
2	2 background	1,05	1,05	1,07	1,03	1,05	1,05

	of the base						
3	Base edge	1,03	1,05	1,02	1,04	1,06	1,04
4	Background duck	1,08	1,10	1,06	1,06	1,10	1,08
5	Weft pattern	307	319	341	328	320	323
6	Weft edge	12,5	13,0	12,8	13,2	12,5	12,8

Table 3.7

Numerical characteristics of the yarn count in the sliver

№	Name	Indicators				
		Dispersion	Mean square deviation *S*	Coefficient of variation *C*	Absolute error *ε*	Relative error *δ*
1	1 base background	0,00025	0,016	1,52	0,02	1,09
2	2 background of the base	0,00020	0,014	0,33	0,02	1,60
3	Base edge	0,00025	0,016	1,54	0,02	1,90
4	Background duck	0,00040	0,020	1,85	0,03	2,30
5	Weft pattern	157,50	12,55	3,90	16,0	4,80
6	Weft edge	0,0950	0,310	2,42	0,40	3,00

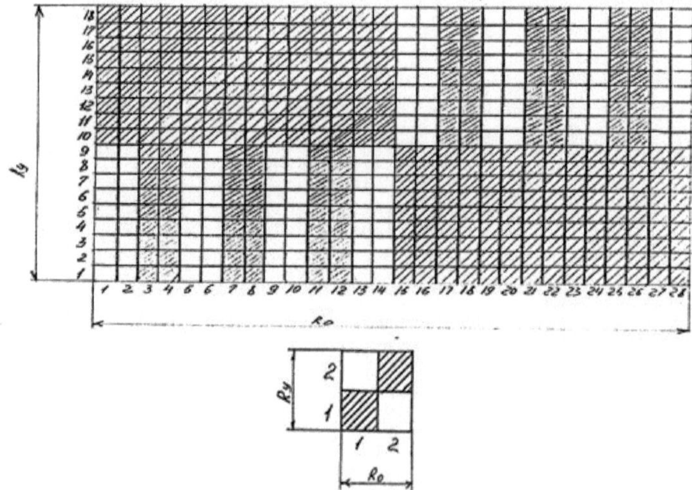

Fig. 3.16. A fragment of a weave pattern on a jacquard ribbon.

Finishing of the embossed weft (according to figure 3.17) of one weave pattern

$$A_y = \frac{AB+BD+DF-AC-CE-EF}{AB+BD+DF} = \frac{2(AB-AC)}{2AB-BD} \qquad (1)$$

Also from Fig. 3.17 it follows. $AC=EF=b/Kn_o$, $BC=DE=b$.

$$AB=\sqrt{BC^2+AC^2}=\sqrt{b^2+(b/K_{Ho})^2} \qquad BD=(R_O+t_y)\ b/K_{Ho}$$

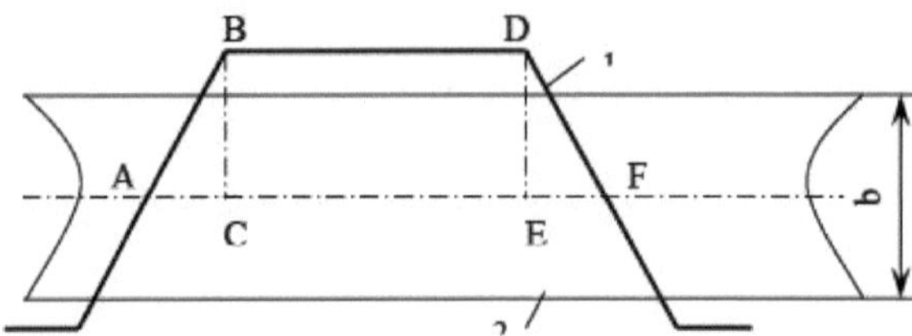

Fig. 3.17. Calculation of relief weft processing, where 1 is relief weft, 2 is background fabric.

In equation (1), the number two means the number of weft transitions from the wrong side to the front side of the fabric, so we can denote by t_y and the formula will take the form after the statement;

$$a_y = \frac{t_y\left(\sqrt{b^2+(b/K_{Ho})^2}-b/K_{Ho}\right)}{t_y\sqrt{b^2+(b/K_{Ho})^2}+(R_o-t_y)\dfrac{b}{K_{Ho}}} * 100 \qquad (2)$$

Equation 2 takes into account the workmanship of the embossed weft for one yarn in a weave pattern. To determine the workmanship in the pattern rapport, it is necessary to take into account the number of repeats of warp and weft rapports in the pattern rapport in the width and length of the ribbon, i.e. *no / Ro+ (ny / Ry) * Ci* (3), where: *Po* - the number of warp yarns in the pattern rapport along the width of the ribbon; *Pu - the number of* weft yarns in the pattern rapport, and along the length of the ribbon; *Ro, Ry - the* rapport of the basic weave of the patterned weft on the warp and weft respectively; *Ci - the* number of relief wefts laid in one shed. Substituting (3) into (2), we have the workmanship of the relief weft in the jacquard ribbon taking into account the pattern report of the fabric.

$$a_y = \frac{t_y(\sqrt{b^2+(b/K_{Ho})^2} - b/K_{Ho}) * (\frac{n_o}{R_o} + C_1\frac{n_y}{R_y}) * 100}{t_y\sqrt{b^2+(b/K_{Ho})^2} + (R_0 - t_y) * b/K_{Ho}} \quad (4)$$

Here is an example of calculation of pattern weft weaving: *b = 0.41 mm; do = 0.144 mm ; dy = 0.12 mm; Ci = 2* - for rapier machine *; KHO = 0.5 ;* n_o = *352.*

yarns; $n_y = 730$; $R_o = 28$; $R_y = 14$; $t_y = 8$ (see Fig. 3.16) for one of a fragment of the weave pattern.

$$A_y = \frac{8(\sqrt{0.41^2+(0.41/0.5)^2} - 0.41/0.5) * (\frac{352}{28} + 2 * \frac{730}{18}) * 100}{8\sqrt{0.41^2+(0.41/0.5)^2} + (28-8) * \frac{0.41}{0.5}} = 307$$

A comparison of the calculated value with the experimental values (Table 3.7) shows that the deviation is about 5 %, which is acceptable in the textile industry. Reducing the thread breakage level in the weaving process contributes to the achievement of the goal. We have adopted the basic thread breakage as a criterion for optimising the weaving process. Also the following basic and independent parameters were chosen: x 1- warp tension, cN; x_2 - the size of the scalp, mm; x_3 - the position of the scalp relative to the sternum in height, mm. It is established that warp thread breakage at small values of the filling tension increases due to the increase of the surge strip. At high values of the filling tension, the thread breakage also increases due to overstressing of the warp threads. When the machine is operated with a larger scalp, the conditions of filling yarn surfing are improved and the size of the surfing strip is reduced, which worsens the conditions of fabric formation and leads to overstressing of warp yarns at surfing, resulting in higher thread breakage. The height position of the scalo determines the equal tension or different tension of the shed branches when the weft is run in to the fabric edge and when the weft is inserted into the shed. The selected factors are not interchangeable, can be measured and changed in minimum and maximum values. Other technological parameters of loom threading, were constant during the experiment. Since the weaving process is non-stationary in time, and when conducting a large number of experiments there is a distortion of the results due to different disturbances of the process, the randomisation of experiments was applied in the planning of the experiment. The control of main yarns breakage was carried out according to the known methodology. The central composite method of experiment planning of the second order was adopted in the work, which gives the possibility of detailed study, description and optimisation of weaving process in the investigated area of optimisation. The selection of intervals and values of factors for five levels of variation was carried out taking into account the technological possibilities of machine filling according to Table 3.8.

Table 3.8.

Levels of variation of factors

Factors	Levels of variation					Interval
	-1,68	-1,0	0	+1,0	+1,68	
x_1 - filling warp tension, cN	13	16	20	24	27	4
X2 - value of the offset, mm	7	10	15	20	23	5
xz - position of the rock relative to the sternum, mm	-15	-10	0	+10	+25	15

The experiment performed on the selected matrix allows us to obtain the a second-order mathematical model describing the influence of factors Xi, X2, xz on the selected optimisation parameters of the following form

$$y = в_0 + в_1x_1 + в_2x_2 + в_3x_3 + в_{12}x_1x_2 + в_{23}x_2x_3 + в_{13}x_1x_3 + в_{11}x_1^2 + в_{22}x_2^2 + в_{33}x_3^2$$

where *vo, ei, 6ij, vi* - regression coefficient; *vo* - free term; *ei* =1, 2, 3 - regression coefficients at linear terms; *ej=1*, 2, 3 - coefficients at interaction of factors; *vc* = - regression coefficients of squared terms.

The paper contains calculations of regression coefficients, dispersion of regression coefficients, Student's and Fisher's criteria. Comparison of calculated and tabulated values of regression coefficients showed that all coefficients are significant and the model is adequate, the mathematical model describing the dependence of breakage on the selected factors has the following form

$$\begin{aligned} \acute{O}_R &= 0{,}113 - 0{,}033X_1 - 0{,}008X_2 - 0{,}011X_3 - 0{,}015X_1 \cdot X_2 + 0{,}023X_1 \cdot X_3 + \\ &+ 0{,}025X_2 \cdot X_3 + 0{,}084X_1^2 + 0{,}018X_2^2 - 0{,}017X_3^2 \end{aligned} \quad (3.1)$$

It is reasonable to evaluate a technology experiment using cut-offs :

- $y = f(x_1)$ at constant x_2, xz;
- $y = f(x_2)$ at constant xi,x_3 ;
- y= f (xz) at constant x1, x .2

Tables 3.10- 3.12 summarise the results of the breakage calculations from the input factors. As can be seen, all equations represent the equation of a parabola. The analysis of curves (Figs. 3.18- 3.21) plotted by the obtained equations shows that the change of y from x_1 and x_2 has the form of concave parabolas (Figs. 3.18, 3.19, 3.21). The influence of xz (position of the rock relative to the breast of the loom Fig. 3.20) on y is represented by a convex parabola with minimum y values at xz equal to - 1.68 and + 1.68 respectively. Consequently, when producing this fabric, it is advisable to raise or lower the scala as much as possible in relation to the sternum, and raising the scala (xz= +1.68), as shown in Fig. 3.21 leads to the lowest breakage. Separately plotted curve of change of breakage y from the position of the offset x_2 at zero value x_1 (tension of the main threads) and the maximum lifted rock xz= +1,68 shows (see Fig. 3.21) that at $x_2 = 0$ it is possible to reduce the breakage of threads in 2.2 times, that is, the parameters will have the following values: tension of the main threads - 20 cN (per 1 thread); the value of the offset - 15 mm.; the position of the rock relative to the sternum - (+ 25) mm. At these values of parameters breakage of main threads will not exceed 0,05 breaks per 1m.

Table 3.10.

Calculation results y=y^x_1) at constant x_2 and xz

№	Constant values of factors	Yarn breakage of warp *y* values of factor x_1 variables				
		-1,682	-1	0	+1	+1,682
1	x_2=-1,xz=-1	0,465	0,283	0,158	0,201	0,327
2	x_2=-1, x=0_3	0,407	0,241	0,139	0,205	0,347
3	x_2=-1,x=1_3	0,261	0,164	0,085	0,174	0,331
4	x_2=0,x=-1_3	0,385	0,247	0,107	0,135	0,251
5	x_2=0, x=0_3	0,407	0,23	0,113	0,164	0,295
6	x_2=0, x=1_3	0,340	0,179	0,085	0,159	0,306
7	x_2=1,x=-1_3	0,449	0,247	0,092	0,105	0,211
8	x_2=1, x=0_3	0,391	0,225	0,123	0,189	0,331
9	x_2=1,x=1_3	0,400	0,229	0,12	0,179	0,316

Table 3.11.

Calculation results y=f(xi) at constant x1 and xz

№	Constant values of factors	Yarn breakage warp variables factor value x_2				
		-1,68	-1	0	+1	+1,68
1	X1=-1,x =-1_3	0,328	0,283	0,247	0,247	0,268
2	X1=-1, x =0_3	0,269	0,241	0,230	0,255	0,293
3	X1=-1,x =1_3	0,176	0,165	0,179	0,229	0,284
4	X1=0,x =-1_3	0,214	0,158	0,107	0,092	0,102
5	X1=0, x =0_3	0,177	0,139	0,113	0,123	0,151
6	X1=0, x =1_3	0,107	0,086	0,085	0,12	0,165
7	X1=1,x =-1_3	0,313	0,247	0,181	0,151	0,151
8	X1=1, x =0_3	0,254	0,205	0,164	0,159	0,176
9	X1=1,x =1_3	0,207	0,175	0,159	0,179	0,213
10	X1=0,x_3 =1.68	0,241	0,115	0,046	0,098	0,155

Table 3.12.

Results of calculation y=Dxz) at constant x_1 and x_2

№	Constant values of factors	Warp breakage *y* variables factor value xz				
		-1,68	-1	0	+1	+1,68
1	X1=-1,x =-1_2	0,190	0,283	0,241	0,165	0,094
2	X1=-1, x =0_2	0,239	0,247	0,23	0,179	0,125
3	X1=-1,x =1_2	0,222	0,247	0,255	0,229	0,192
4	X1=0,x =-1_2	0,151	0,158	0,139	0,086	0,051
5	X1=0, x =0_2	0,115	0,107	0,103	0,085	0,077
6	X1=0, x =1_2	0,052	0,092	0,123	0,120	0,098
7	X1=1,x =-1_2	0,179	0,201	0,205	0,175	0,135
8	X1=1, x =0_2	0,096	0,135	0,164	0,159	0,136
9	X1=1,x =1_2	0,06	0,100	0,159	0,179	0,173

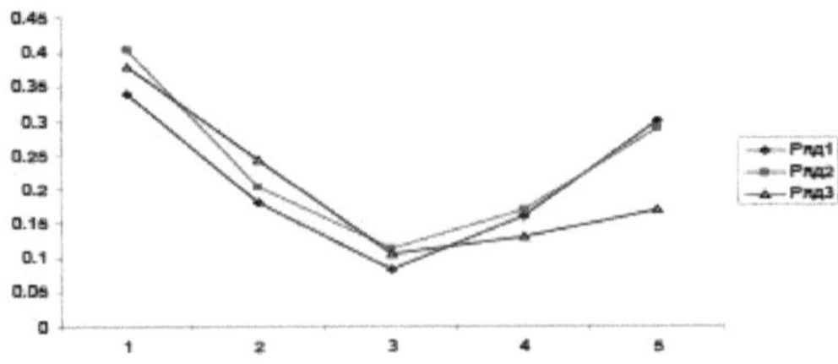

Row1 - X2 = 0 ; xz = 1 Row2- X2 = 0 ; xz= O RowZ - X2 = 0 ; xz = -1
Fig. 3.18 Influence of warp filling tension on thread breakage

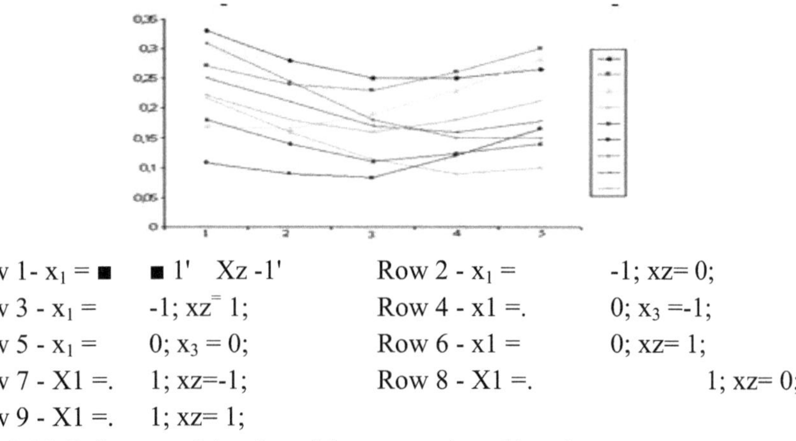

Row 1 - x_1 = ■ ■ 1' Xz -1' Row 2 - x_1 = -1; xz= 0;
Row 3 - x_1 = -1; xz= 1; Row 4 - x1 =. 0; x_3 =-1;
Row 5 - x_1 = 0; x_3 = 0; Row 6 - x1 = 0; xz= 1;
Row 7 - X1 =. 1; xz=-1; Row 8 - X1 =. 1; xz= 0;
Row 9 - X1 =. 1; xz= 1;

Fig. 3.19. Influence of the size of the gap on thread breakage

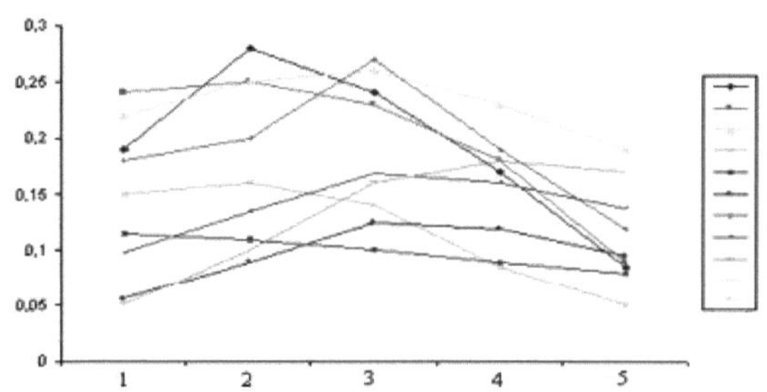

Row1-X1 = -1; x_2 =-1;Row2-X1 = -1; x_2 = 0;
RowZ-X1 = -1; x_2 =1; Row4-X1=0 ; x_2 =-1;
Row5-X1= 0; x_2 =0; Row 6 - X1 =0 ; x_2 = 1;
Row7-X1=1; x_2 = -1; Row8-X1=1; x =0;2
Row 9 -X1 =1; x_2 = 1;

Figure 3.20: Effect of scalo position on thread breakage

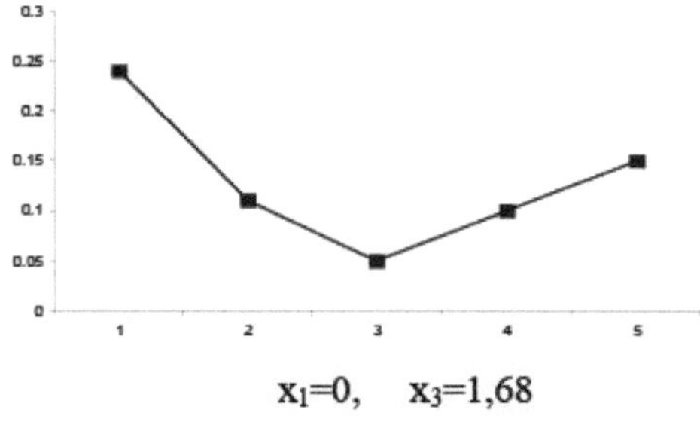

x_1=0, x_3=1,68

Fig. 3.21. Influence of the size of the gap on thread breakage

In the abstract, we note. The method of fixing sagging patterned weft yarns on the wrong side and enhancing the relief effect on the front side of jacquard tape is proposed. The technological sequence of designing, preparation of jacquard ribbons special purpose production is developed. It is expedient to produce special-purpose jacquard ribbons on the basis of local raw materials (natural silk), as this type of raw materials is the most available in the Republic of Uzbekistan in the required quantity and price and has good ergonomic properties. The formula of relief weft processing for two-stitch jacquard fabrics with a large pattern spread is obtained. The density on the basis, width and number of ends in the threading when producing jacquard ribbons are substantiated. The technological process of jacquard ribbon production on the weaving machine by means of the mathematical method of second-order rototable experiment planning is investigated. The geometrical interpretation of the mathematical model is studied by means of slices. The optimum technological parameters of jacquard sliver production are determined. The level of warp thread breakage is 0.05 breaks per 1 m of fabric at warp filling tension of 20 cN, the value of backstitch -15 mm. and the position of the scalo above the sternum by 25 mm.

In the fourth chapter the warp tension and consumer and physical-mechanical properties of jacquard ribbons are investigated. Rank evaluation of exploitation properties of special purpose jacquard ribbons was carried out and the quality of jacquard ribbons was investigated. The developed jacquard tapes have a special purpose, which makes specific requirements to them. These are the requirements of the consumer, technological possibilities of production, and most importantly - a set of properties that ensure the use of their purpose. Besides, it is necessary to fulfil aesthetic requirements. Usually bulkiness and relief are given to ribbons by large-patterned weaves with elongated overlaps, which makes it difficult to bind them to clothes. All these requirements have been achieved by creating jacquard ribbons of new types of weaves, which, by reducing the length of overlaps, allow to obtain a more rigid structure. In addition, this type of weave allows to obtain geometric patterns of ornamentation of the main field of epaulettes, which reflect the heritage of ancient monuments and subtleties of aesthetic tastes of the peoples of Uzbekistan. Polyester yarns of different linear density were used for manufacturing of ribbons. The characteristics of warp and weft yarns are given in Table 4.1. The obtained results showed that one type of warp yarns was used to produce jacquard ribbons of all variants, and only weft yarns of different linear density were varied. This made it possible to obtain ribbons with different structures. It is obvious that the developed new assortments of jacquard tapes have no analogues. Their main characteristics are given in Table 4.2. The obtained results showed that all samples of jacquard tapes have the same density parameters. However, due to the use of warp and weft yarns of different thicknesses, tapes with different surface densities were obtained

Table 4.1.

Characteristics of yarns and yarns for the production of jacquard ribbons

Type of sample	Assignment					
	Basis			Duck		
	Fibre composition	Linear Density T	Cr\m spinning	Fibre composition	Linear Density T	Cr\m spinning
1	PEF	11,0	600	PEF	8,7	120
2	PEF	11,0	600	PEF	16,6	
3	PEF	11,0	600	PEF	5,7	120

Characteristics of jacquard ribbons

Table 4.2.

No. of Obr	Type of binding	Fibre composition linear density		Density per 10cm.		Surface density g/cm2
		basis	ducks	bases a	Duck	
1	Jacquard	PEF	PEF	60	25x2	13,78
		11 tex	8,7 T			
		600 kr\m	120 kr\m			
2	Jacquard	PEF	viscose	60	25x2	13,84
			16,6T			
3	Jacquard	PEF	PEF	60	25x2	13,81
			5,7T			
			120 kr\m			
4	Jacquard	PEF	PEF	60	25x2	13,83

Specific operating conditions require special properties, which were identified by the method of expert judgement by ranking the properties according to their degree of importance. The results of expert evaluations showed that the most significant for these operating conditions are: elongation at break; stiffness; extensibility; breaking load. Undoubtedly significant are: appearance; comfort. The most significant are mechanical properties that provide technological properties to the products made of the developed jacquard tapes. During their production the tapes experience biaxial spatial tension in two mutually perpendicular directions. The deformation of the tapes has a complex character and is accompanied by displacement of structural elements and stretching. According to the ranking of the quality indicators, the first six for the jacquard weave assortments were tested. Physical, mechanical and other properties of the tapes were investigated according to the current methods of the Republican Standards. The tests were carried out on the samples produced in the conditions of the production laboratory of JSC "Uchkun" of the State Association of Uzbekistan, "Centexuz" certification laboratory. The main characteristics of tapes are estimated by a complex of parameters determining physical-mechanical, aesthetic, hygienic and operational properties of fabrics, and for some of them there are objective instrumental methods of estimation and instrumental support, and for others, which are not subject to such estimation, an individual approach of an expert is used. The non-measurable criteria include such characteristics as appearance and comfort. Appearance in the experimental studies was assessed by experts visually by a set of characteristics - pattern geometry (design), shine (satin effect), colour, resistance to light and wet treatments. Of the above parameters, the first two were evaluated by experts and were highly appreciated for their clear relief pattern. Comfortability is assessed by a complex of characteristics determining hygroscopicity, thermal insulation properties, air permeability and other parameters. Since the comfort criterion is very subjective, it was evaluated based on the results of a trial of 20 sets of military epaulettes. According to the specialists' feedback, the product is characterised by high heat-protective properties that create comfortable conditions. According to this feature and the opinion of specialists, silk ribbons meet the conditions of operation in areas with hot climate. During production and operation jacquard ribbons are subjected to various mechanical effects causing tensile, compression, bending deformations, as well as tangential resistance of fabrics and also associated with such phenomena as sliding and crumbling. Tensile strength is an important indicator of the mechanical properties of textile fabrics, including jacquard ribbons, which determine their integrity. Tensile strength is assessed by the tensile load - the highest force withstood by the sample at the moment of rupture. One of the most important characteristics determining the relation of tapes to various applied loads and its integrity is breaking load, it characterises the strength of tapes under biaxial static tension. The results of determination of breaking load and elongation at break are given in Table 3. The obtained results showed that all the samples under study have close values of load and elongation at break, which indicates the identity of their structure. At rigid fastening of jacquard tapes in the form of epaulettes on clothes they are constantly subjected to tensile biaxial stresses, which causes displacement of yarns of one system relative to the other, i.e. sliding. Under incomplete design is understood the autonomous use of separate CAD subsystems, where the values of the coefficient of connectivity or tension of fabric production on the weaving machine are detected, according to which the warp and weft yarn densities necessary for this weave and the accepted type of raw material are determined. Oriented design involves the development of a fabric prototype.

Shearability is measured by the amount of load required to displace the warp or weft yarns. The shear resistance of yarns is determined by many factors of fibre and yarn structure, type of weave, yarn density, warp and weft thickness ratio, yarn bending, finishing features, etc., which influence the amount of friction between the fibres. The results of determining the extensibility showed that all variants of the developed tapes are not extensible (the norm of extensibility is not less than 9 kg). Resistance to wet processing - the property of preserving the tone of the dyed fabric and the transition of dye on white fabric and other colours, as well as the preservation of linear dimensions (shrinkage). In this work for evaluation of criteria of properties of combined tapes the method of a priori ranking according to the method of planning of experiments is used. During operation, epaulet ribbons are constantly exposed to abrasion, which causes their wear and pilling, i.e. their appearance deteriorates. The results of rubbing wear are summarised in Table 4.3. The obtained results showed that all variants of jacquard tapes are wear-resistant and do not form pilling on the surface. At the same time, they do not lose colour when rubbed in wet condition.

Table 4.3.

Results of determination of physical and mechanical properties of jacquard ribbons

No. Obr.	Breakin g load	Elongatio n	Wear Cycle.	Conditiona l stiffness µN cm2	The durabilit y of the colouring	Extensibilit y Kg.	Pilen-durability of saws\10cm
	H	%					

					ball.		2
1	590,2	34,4	19750	9800	5	18,3	1
2	617,6	35,2	20080	10300	5	20,4	1
3	575,8	34,1	18540	8980	5	19,8	1
4	589,8	33,8	19720	9050	5	17,3	1
5	550,3	32,0	20020	9930	5	18,7	1
6	580,7	34,0	19870	8720	5	18,8	1

Table 4.4.

Quality parameters of jacquard ribbon selected as a result of ranking assessment

Jacquard ribbon (epaulet) made of yarns	
Base polyester, weft polyester	Base-polyester, weft-natural silk.
xZ - appearance	*xZ* appearance
x2 comfort	*x2* comfort
x8- elongation	*x8*- elongation
cotton - stiffness	*cotton* - stiffness
x4 extensibility	*x4* extensibility
x7 - breaking load	*x7* - breaking load

In terms of abrasion resistance, the experimental samples are practically as good as silk fabrics. Abrasion was measured at M -235/3.

Samples	Unit. Measurements	magnitude
1 option	cycle	7000
2 variant	cycle	3950
3 option	cycle	7005
The average is 6,985 cycles.		

The breaking load was determined at AG - 1. The kit includes a personal computer, printer and compressor.

BasisDuck

Fo = greater than 1019NF_y = 624N

Eo = greater than 18%Eu = 18%

According to the results of research of physical-mechanical and technological properties of new assortments of jacquard fabrics the following main conclusions can be made: among numerous parameters of new assortments of jacquard weave fabrics the priority are appearance and resistance to wet processing; according to physical-mechanical and hygienic properties new fabric samples meet the requirements of the Republican and European standards.

CONCLUSION

At the present stage, the development of military attributes (epaulettes, chevrons, etc.) for the Armed Forces of the Republic of Uzbekistan, taking into account national traditions in ornamentation and symbolism is one of the most important tasks in the political sphere, so the development of technology and design of attributes for the Armed Forces of the Republic of Uzbekistan is a necessary aspect in which it is necessary to take into account the semantic concepts of "History-modernity". Computer design of the machine causes mainly expansion of assortment of produced products and more convenient storage of information and weave pattern, and the artist-dissenator can on the monitor without working out samples apply multivariation approach of production of the given pattern, apply different combinations of tones and weaves and transfer the programme into the computer disk, controlling electronic jacquard machine, that would require much time and big material expenses at development of pattern by classical way. Jacquard ribbons for epaulettes have been developed, taking into account new ornamental compositions and colour solutions in accordance with symbolic meanings between them. New systems of jacquard ribbons pattern coding are proposed, taking into account the weaves of ground and patterned yarns, the type of jacquard machine and the sequence of used weaves. The technological sequence of designing, preparation and production of special purpose jacquard ribbons is developed, the method of fixing sagging patterned weft threads on the wrong side of the ribbon and enhancing the effect of relief on the front side of the jacquard ribbon is specified. It is expedient to produce special purpose jacquard ribbons on the basis of local raw materials (natural silk), as this type of raw materials is available in the required quantity and price and has ergonomic properties. The formula of weft processing for two-weft jacquard fabrics with a large pattern rapport is obtained. The filling density on the base, width and number of ends in the filling when producing jacquard ribbons are substantiated. The technological process of jacquard ribbon production on the weaving machine is investigated by means of the mathematical method of second-order rototable experiment planning. The geometric interpretation of the mathematical model is studied by means of slices. The optimum technological parameters of jacquard sliver production are determined. The level of warp thread breakage is 0.05 breaks per 1m. of fabric at warp tension-20 cN, the value of scoring-15 mm, the position of the scalp above the sternum by 25 mm. In the assortment of jacquard weave tapes the priority parameters are appearance, comfort, breathability, non wrinkling, breaking load and resistance to wet processing. In terms of physical, mechanical and hygienic properties new samples of jacquard tapes meet the requirements of the Republican and European standards for the group of special purpose jacquard tapes.

LITERATURE

1. Ogarkov N.V. Military Encyclopaedic Dictionary. -M.: Military Publishing House, 1983. -863 c.
2. Pokhlebkin V. V. International symbols and emblematics. -M.: International Relations, 1989. -300 c.
3. Karamatov X., Rtveladze E., Saidkasimov S. Amir Timur in World History. -T.: Shark, 2001. -174 c.
4. http:/army, armor.kiev.ua/forma-2/index- shtml
5. http : /uzshevron, nm, ru./pogqk. htm
6. Maraimov C.T. et al. Shoulder insignia. Patent SPD 00850. 2001. №3.
7. Maraimov S.T. et al. Shoulder insignia. Patent SPD 00970. 2002. №5.
8. Uzakova U.R. et al. Shoulder insignia. Patent. 2003. №.
9. Malakhova S.A., Zhuravleva T.A. Artistic design of products. -M.: Legprombytizdat, 1988, - 304 p.
10. Sobolev P.P. Essays on the history of fabric decoration. Academia Moscow - St. Petersburg. 1934,- 434 c.
11. Martynova A.A. et al. Structure and design of tissues. -M.: RIO MGTOA, 1999. - 434 c.
12. Nikitin NM. Theory of weaving weaves on a mathematical basis. -M.: Light Industry, 1964. - 455 c.
13. Nikitin N.M. Designing of fabrics. -M.: Rostekhizdat, 1961. - 212 c.
14. Kalistratov M.A. Automation of designing and analysis of weaving weaves by means of computer. Avtoref. Cand. Cand. of Technical Sciences. - MOSCOW INSTITUTE OF TECHNOLOGY. 1969. - 16 c.
15. Gleaner M.I. et al. Methods of pattern programming for double fabrics. T.T.P., Izvestiya Vuzov, 1971. -№4, - 89 c.
16. Milasius V.M., Reklaitis V.K. Coding of weaving weaves. -M.: Legprombytizdat. 1988. - 80 c.
17. Zhuravleva T.A. Artistic design of industrial fabric patterns with the use of computers. Autorref. dne. ... Cand. of technical sciences. -M.: MTI, 1984. - 17 c.
18. Borzunov G.I. Computer application in preparation for production of fine-pattern weave fabrics on STB machines. Autorref. dne. ... Candidate of Technical Sciences. -M.: 1980. - 17 c.
19. Borzunov G.I. Analysis of coloured weaving patterns by means of computer. T.T.P., IzvestiyaVuzov, 1985. -№5. -C. 35-38.
20. Krylov G.A. Automation of design of patterns of crepe-type weave by

means of computer. Avtoref. dne. ... Cand. of Technical Sciences. -M.: TSNIIHBI, 1983. - 22 c.
21. Alimbaev E.Sh. Tutsima tuzilishi nazariyasi. -T.: Alokachi, 2005. -231 б.
22. Alimbaev E.Sh. Tukuv urilishlarni tasniflash. -T.: TTESI, 1996. - 18 б.
23. Daminov A.D. Fundamentals of structure forecasting and design of textile fabrics. Autorref. dne. ... doctor of technical sciences. -T.: TITLP,2006. -42 c.
24. Surnina N.F. et al. Automation of designing of fabrics. Textile industry. 1989. -№9. -C. 60.
25. Lomov S.V., Gusakov A.V. Coding of weave of layer-frame woven structures. Izvestiya Vuzov. Technology of textile industry, 1993. -№3. - c. 43-50.
26. Onikov E.A. et al. Reference book "Cotton weaving". -M.: Light Industry, 1979. - 488 c.
27. Surnina N.F. Designing fabric according to the given parameters. - M.: Light Industry, 1973. - 144 c.
28. Damyanov G.B. et al. Fabric structure and modern methods of its design. -M.: Light and Food Industry, 1984. -237 c.
29. Smirnov V.I. Theoretical studies of the structure of plain weave fabric. -M.: Rostekhizdat, I960. - 100 c.
30. Popovskiy A.U. Study of changes in qualitative indicators of natural silk threads during processing into aurora fabrics. Avtoref. dne. ... Cand. of technical sciences. -T.: TITLP, 1977. -27 c.
31. Alekseev K.G. Research of the process of formation of cotton fabric of plain weave. -M.: Gizlegprom, 1958. - 146 c.
32. Ilyin I.V. About geometrical structure of single-layer fabric. Izvestiya Vuzov. T.T.P., I960. 35, c. 61-66.
33. Vorobyev V.A. Calculation methods for construction of woollen yarn and fabric. -M.: Light Industry, 1964. - 163 c.
34. Kutepov O.S. Structure and design of fabrics. -M.: Legprombytizdat, 1988. -220 c.
35. Malakhova S.A. Artistic design of textile products. -M.: Legprombytizdat, 1988. - 303 c.
36. Kozlov V.N. Fundamentals of artistic design of textile products. -M.: Legprombytizdat, 1981. -264 c.
37. Zaborovskiy B.A. et al. Automatic programming of jacquard patterns. -K.: 1978. - 135 c.
38. http://ecoguild.narod.ru
39. Dekhanova M.G., Mshvenieradze A.P. Ribbon weaving and wicker

production. -M.: Legprombytizdat, 1987. -200 c.
40. Dekhanova M.G. Ribbon weaving and wicker equipment :
Textile and haberdashery industry, 1989. - 99 c.
41. http://mueller-frick.com
42. Beresneva V.Ya., Romanova N.V. Questions of ornamentation of fabrics. -M.: Light Industry, 1977. - 192c.
43. Uzakova U.R., Alimova H.A., Rakhimkhodjaev S.S. Fundamentals of the development of elements of attributes of the armed forces. // Respublika ilmiy-amaliy anjuman. - Toshkent., 2005. - 221-226 б.
44. Uzakova U.R., Alimova H.A., Rakhimkhodjaev S.S. Fundamentals of design of military attributes for the armed forces of the Republic of Uzbekistan. // -Tashkent, 2005. -№ 2, - C. 115-127.
45. Rosenblum E. The artist in design. -M.: 1975. - 368 c.
46. http://www.design.ru
47. Uzakova U.R., Daminov A.D., Zholdasova A.D. Classical ornament in Karakalpak clothing. // Abstracts of the scientific and practical conference "Tutsimachilik 2005". - Tashkent, 2005. -C.81.
48. Uzakova U.R., Rakhimkhodjaev S.S., Surova E. Design and design of tapestry fabrics. // Abstracts of the scientific and practical conference. - Tashkent, 2004. -C. 103-104.
49. Uzakova U.R., Khasanov B.K., Khakimova M.Sh. Design and technology of carpet products for children's interiors. // Theses of reports of the scientific and practical conference. -Tashkent, 2008. - C 88-91.
50. Mirzaev O.A., Uzakova U.R., Daminov A.D. Modelling and design of fabric with the help of computer graphics. // Collection of materials "Search 2004". -Ivanovo, IGTA. 2004. - C. 188-190.
51. Sobirova G.N., Rakhimkhodjaev S.S., Uzakova U.R. Software of the subsystem of weave weave construction. // Problems of textile, 2005. -№2- c. 62-64.
52. Uzakova U.R., Rakhimkhodjaev S.S., Sobirova G.N. Software of the subsystem of weave weave construction. // Theses of reports of the scientific and practical conference. -Tashkent, 2005. - C 80.
53. Uzakova U.R., Khasanov B.K., Kadyrova M.Y. Technology and design of the fabric of children's assortment. // Theses of reports of the scientific-practical conference. -Tashkent, 2008. - C 92-95.
54. Surova E.A., Uzakova U.R., Rakhimkhodjaev S.S. Tapestry fabric of a new structure. // Collection of materials "Search - 2004". Ivanovo, IGTA, 2004. - C. 103-104.

55. Uzakova U.R., Shekerova S. Obtaining the finished semi-finished product from the jacquard machine. // Theses of reports of the scientific-practical conference. - Tashkent, 2004. - C 81.
56. Saimuratov H.S., Uzakova U.R. Code description and construction of weaves of jacquard fabrics. // Republic ilmiy-amaliy conference "Tukimachilik-2004y.", -Toshkent, 2004. -98 б.
57. Uzakova U.R., Alimova H.A., Rakhimkhodjaev S.S. Jacquard complex in weaving on the electronic basis. // NTZH. Ipak. 2002. -№1, -C. 34-35.
58. Rakhimkhodjaev S.S., Uzakova U.R., Kadyrova M.A. Computer technologii yerdamida jakkard tuktsmalarni patronlashtirish. 1 va 2 kiem. -T.: 2007.-274 b.
59. Uzakova U.R., Alimova X. A., Rakhimkhodjaev S.S. Jacquard complex in weaving on electronic basis. Information leaflet of GFNTI, -Tashkent, 2001.
60. Uzakova U.R., Alimova H.A., Rakhimkhodjaev S.S. Development and production of woven tapes. // Material of the international scientific and technical conference. -Tashkent, 2002, -C. 21.
61. Granovsky S.G. Jacquard fabrics (pattern patronisation) - M.: Legkaya Industriya, 1971. - 368 c.
62. Uzakova U.R., Alimova H.A., Rakhimkhodjaev S.S. Jacquard ribbons for special purposes. // N.T.J. Silk. -Tashkent, 2002. -№1, -C.4950.
63. Rozanov F.M. et al. Structure and design of fabrics. -M.: Light Industry, 1953. -471 c.
64. Sevastyanov A.G. Methods and means of research of mechanical and technological processes of textile industry. -M.: Light Industry, 1980. - 392 c.
65. Getsonok B.I. Statistical control of weaving process. -M.: Light and food industry, 1983. - 86 c.
66. Shutova N.E. Thread breakage and stability of technological process. -M.: Light Industry, 1975. - 79 c.
67. Uzakova U.R., Rakhimkhodjaev S.S. Optimisation of technological process of weaving at production of jacquard ribbons. // Materials of scientific and practical conference "Development and improvement of design and technology of leather goods", -Tashkent, 2008.-C. 167-171.
68. Solov'ev A.N., Kiryukhin S.M. Quality assessment and standardisation of textile materials. -M.: Legkaya Industriya, 1974.-245 pp.
69. Chernenko N.A., Uzakova U.R. Designing decorative fabric on "Riomo SISTEMS". // Respublika ilmiy-amaliy conference "Tutsimachilik-2004y.". - Toshkent, 2004. -74 б.

70. Shamukhitdinova L.Sh., Chursina V.A., Uzakova U.R., Alimova H.A. Development of the principles of creating new types of weaving patterns for jacquard fabrics for clothing purposes. // Problems of textile. -Tashkent, 2003. -№3, - c. 5-7.
71. Manokhina O., Daminov A.D., Development of weaving weave pattern programme for implementation on Jacquard machine. // Collection of scientific papers of TDTU. -Tashkent, 2003. - 160-161 c.
72. Uzakova U.R., Daminov A.D., Abdurazokov Sh. Sport fabric for wrestlers "Kurash". // Theses. dokl. of the scientific-practical conference, - Tashkent, 2004. - c. 86

Printed by Books on Demand GmbH, Norderstedt / Germany